DEFENSE, SECURITY AND STRATEGIES

FEDERAL PROTECTIVE SERVICE

ASSESSMENTS AND FEDERAL FACILITY SECURITY

DEFENSE, SECURITY AND STRATEGIES

Additional books in this series can be found on Nova's website under the Series tab.

Additional E-books in this series can be found on Nova's website under the E-book tab.

GOVERNMENT PROCEDURES AND OPERATIONS

Additional books in this series can be found on Nova's website under the Series tab.

Additional E-books in this series can be found on Nova's website under the E-book tab.

DEFENSE, SECURITY AND STRATEGIES

FEDERAL PROTECTIVE SERVICE

ASSESSMENTS AND FEDERAL FACILITY SECURITY

EMILIUS M. ZAMACO
EDITOR

NOVINKA

New York

Library of Congress Cataloging-in-Publication Data

ISBN: 978-1-62417-205-2

Published by Nova Science Publishers, Inc. † New York

CONTENTS

PREFACE

The security of federal government buildings and court facilities affects not only the daily operations of the federal government but also the health, well-being, and safety of federal employees and the public. In the 112th Congress, legislation was introduced in the House of Representatives and Senate to improve the Federal Protective Service (FPS), the agency charged with responsibility to protect federal buildings. This book provides an overview of the FPS whose mission is to protect more than 9,000 Federal buildings and the 1.4 million Federal employees and visitors who occupy them throughout the country every day by leveraging the intelligence and information resources of its network of public and private sector partners.

Chapter 1 - The security of federal government buildings and court facilities affects not only the daily operations of the federal government but also the health, well-being, and safety of federal employees and the public.

Early in the 112th Congress, legislation was introduced in the House of Representatives and Senate to improve the Federal Protective Service (FPS), the agency charged with responsibility to protect federal buildings, the employees who work in the buildings, and public visitors. On January 5, 2011, H.R. 176, the Federal Protective Service Improvement and Accountability Act of 2011, was introduced in the House. On April 8, 2011, similar legislation, S. 772, the Supporting Employee Competency and Updating Readiness Enhancements for Facilities Act of 2011, was introduced in the Senate. The purpose of both bills is to strengthen the security of federal facilities and the ability of FPS to provide the necessary security.

For the purposes of this chapter, federal facilities include any building leased or owned by the General Services Administration. In FY2007, the federal government's real property portfolio comprised 446,000 buildings with an area of 3.3 billion square feet and a replacement value of $772.8 billion. Federal courthouses and facilities are also discussed in this chapter. Additionally, it should be noted that many Members of Congress have state and district offices located in multi-tenant federal buildings.

Security of federal facilities includes physical security assets such as closed-circuit television cameras, barrier material, and security guards (both federally employed and contracted). Federal facility security practices have been subject to criticism by some government auditors and security experts. Elements that have received criticism include the use of private security guards, the management and security practices of the FPS, and the coordination of federal facility security.

Chapter 2 - The Department of Homeland Security's (DHS) Federal Protective Service (FPS) protects over 9,000 federal facilities under the custody and control of the General Services Administration (GSA). In 2007, FPS adopted an inspector-based workforce approach and indicated it would increase its reliance on state and local law enforcement agencies to respond to incidents at these facilities. These facilities range from facilities of proprietary or concurrent jurisdiction— where authority is shared by federal and state and local police—to facilities of exclusive jurisdiction, where only federal law enforcement has authority. As requested, this chapter assesses FPS's efforts to collaborate with state and local law enforcement for assistance in responding to incidents at these federal facilities. GAO reviewed documents on collaboration, GSA and FPS facility data, and GAO's work on key collaboration practices and internal control standards. GAO also contacted 73 selected state and local law enforcement agencies from geographic jurisdictions of varying population sizes and FPS buildings throughout the United States and interviewed FPS and GSA officials.

Chapter 3 - FPS provides security and law enforcement services to over 9,000 federal facilities under the custody and control of the General Services Administration (GSA). GAO has reported that FPS faces challenges providing security services, particularly completing FSAs and managing its contract guard program. To address these challenges, FPS spent about $35 million and 4 years developing RAMP—essentially a risk assessment and contract guard oversight tool. However, RAMP ultimately could not be used because of system problems.

GAO was asked to examine (1) the extent to which FPS is completing risk assessments; (2) the status of FPS's efforts to develop an FSA tool; and (3) FPS's efforts to manage its contract guard workforce. GAO reviewed FPS documents, conducted site visits at 3 of FPS's 11 regions, and interviewed FPS officials and inspectors, guard companies, and 4 risk management experts.

Chapter 4 - This is an edited, reformatted and augmented version of testimony given on July 24, 2012 before the House Committee on Homeland Security, Subcommittee on Cybersecurity, Infrastructure Protection, and Security Technologies.

In: Federal Protective Service
Editor: Emilius M. Zamaco

ISBN: 978-1-62417-205-2

Chapter 1

FEDERAL BUILDING, COURTHOUSE, AND FACILITY SECURITY*

Lorraine H. Tong and Shawn Reese

SUMMARY

The security of federal government buildings and court facilities affects not only the daily operations of the federal government but also the health, well-being, and safety of federal employees and the public.

Early in the 112th Congress, legislation was introduced in the House of Representatives and Senate to improve the Federal Protective Service (FPS), the agency charged with responsibility to protect federal buildings, the employees who work in the buildings, and public visitors. On January 5, 2011, H.R. 176, the Federal Protective Service Improvement and Accountability Act of 2011, was introduced in the House. On April 8, 2011, similar legislation, S. 772, the Supporting Employee Competency and Updating Readiness Enhancements for Facilities Act of 2011, was introduced in the Senate. The purpose of both bills is to strengthen the security of federal facilities and the ability of FPS to provide the necessary security.

For the purposes of this chapter, federal facilities include any building leased or owned by the General Services Administration. In FY2007, the federal government's real property portfolio comprised 446,000 buildings with an area of 3.3 billion square feet and a

* This is an edited, reformatted and augmented version of Congressional Research Service, Publication No. R41138, dated December 9, 2011.

replacement value of $772.8 billion. Federal courthouses and facilities are also discussed in this chapter. Additionally, it should be noted that many Members of Congress have state and district offices located in multi-tenant federal buildings.

Security of federal facilities includes physical security assets such as closed-circuit television cameras, barrier material, and security guards (both federally employed and contracted). Federal facility security practices have been subject to criticism by some government auditors and security experts. Elements that have received criticism include the use of private security guards, the management and security practices of the FPS, and the coordination of federal facility security.

INTRODUCTION

Prior to the April 19, 1995, bombing of the Alfred P. Murrah Building in Oklahoma City, the federal government had no formally established security standards for federally owned or leased facilities.[1] Immediately following the bombing, President William J. Clinton directed the Department of Justice (DOJ) to assess the vulnerability of federal facilities to terrorist attacks or violence and to develop recommendations for minimum security standards.[2] The U.S. Marshals Service (USMS), within DOJ, coordinated two working groups to accomplish these presidential directives. The working groups identified and evaluated various security measures and activities that could address potential vulnerabilities, and minimum security standards were also proposed for federal facilities. Additionally, USMS deputies and General Services Administration (GSA) security specialists conducted inspections at more than 1,200 federal facilities to obtain security data on buildings for use in upgrading existing conditions to comply with the proposed minimum standards. The result of the working groups' efforts was the report *Vulnerability Assessment of Federal Facilities*.[3] This report was significant because it represented the first time that broad security standards were applied to federal facilities.

After the report was issued, President Clinton directed all executive branch agencies to begin upgrading their facilities to meet the recommended minimum security standards. Following the DOJ recommendations, President Clinton also required GSA to establish building security committees for all GSA facilities.[4]

In the 112th Congress, legislation was introduced in the House of Representatives and Senate to improve the Federal Protective Service (FPS), the agency charged with responsibility to protect federal buildings. In the House, H.R. 176, the Federal Protective Service Improvement and Accountability Act of 2011, was introduced on January 5, 2011. Similar legislation, S. 772, the Supporting Employee Competency and Updating Readiness Enhancements for Facilities Act of 2011 (SECURE Facilities Act of 2011) was introduced in the Senate on April 8, 2011. The purpose of both bills is to strengthen the security of federal facilities and improve the safety of employees who work there and public visitors by enhancing the ability of FPS to provide the necessary security.[5] The two bills are discussed in this chapter under the section, "Legislation in 112th Congress: Federal Protective Service."

FEDERAL FACILITY SECURITY LEVELS

Because of the differences among federal buildings and their security needs, USMS categorized federal facilities into five classes based on building size, agency mission and function, tenant population, and the degree of public access to the facility, and developed security standards corresponding to the security level needed for each class:

- Level I—buildings with no more than 2,500 square feet, 10 or fewer federal employees, and limited or no public access;
- Level II—buildings with 2,500 to 80,000 square feet, 11 to 150 federal employees, and moderate public access;
- Level III—buildings with 80,000 to 150,000 square feet, 151 to 450 federal employees, and moderate to high public access;
- Level IV—buildings with 150,000 square feet or more, more than 450 federal employees, and a high level of public access; and
- Level V—buildings that are similar to Level IV but are considered critical to national security (for example, the Pentagon).[6]

INTERAGENCY SECURITY COMMITTEE (ISC)

The Interagency Security Committee was established in 1995, originally as part of GSA, and was transferred to the Department of Homeland Security (DHS) in 2003. The following two sections describe the ISC's work under GSA and DHS chairmanships.

1995-2003: GSA Chairmanship

On October 19, 1995, President Clinton issued an executive order that established the ISC to address "continuing government-wide security" for federal facilities.[7] Chaired at the time by the GSA Administrator, the ISC was composed of representatives from each of the executive branch agencies. The ISC was authorized to consult with other entities, including the Administrative Office of the U.S. Courts, in order to perform its duties. The executive order directed the ISC to

1. establish policies for security in and protection of Federal facilities;
2. develop and evaluate security standards for Federal facilities, develop a strategy for ensuring compliance with such standards, and oversee the implementation of appropriate security measures in Federal facilities; and
3. take such actions as may be necessary to enhance the quality and effectiveness of security and protection of Federal facilities, including but not limited to:
4. encouraging agencies with security responsibilities to share security-related intelligence in a timely and cooperative manner;
5. assessing technology and information systems as a means of providing cost-effective improvements to security in Federal facilities;
6. developing long-term construction standards for those locations with threat levels or missions that require blast resistant structures or other specialized security requirements;
7. evaluating standards for the location of, and special security related to, day care centers in Federal facilities; and
8. assisting the Administrator in developing and maintaining a centralized security data base of all Federal facilities.[8]

Following its establishment, the ISC began to address new security technology developments, cost considerations, and the need to balance security standards with public access to federal buildings. In May 2001, the ISC issued its *Security Design Criteria for New Federal Office Buildings and Major Modernization Projects* (updated in 2004), based on the five security levels for federal facilities.[9] This document required new construction projects to include the use of window glazing protection, establish minimum acceptable distances between federal buildings and streets, control vehicular access to buildings, and evaluate the location and securing of air intake vents.

The September 2001 terrorist attacks on the Pentagon and the World Trade Center heightened concerns about the vulnerability of federal buildings to violence or bombings. In response to these events, the ISC issued revised procedures to respond to potential vehicle bomb attacks by recommending that new federal buildings be constructed at a minimum distance of between 20 to 50 feet from the nearest perimeter barrier, depending upon the security level.[10]

Even though the ISC successfully completed its security design criteria and related documents, a 2002 GAO report found that the committee had made "little progress" in other mandated responsibilities. While GAO reported that the ISC was successfully disseminating security information to member agencies, it also found that the committee's effectiveness was hindered by GSA's "lack of aggressive leadership and support," in that the agency failed to issue operating procedures and did not provide sufficient staff support and funding. GSA was also unable to provide any documentation indicating that the agency or the ISC had actually monitored agency compliance with the security recommendations.[11]

2003-Present: DHS Chairmanship

Congressional enactment of the Homeland Security Act[12] in 2002 and the creation of DHS centralized the federal government's efforts to respond to terrorism, including enhancing physical security for federal facilities. Accordingly, the chairmanship of the ISC was transferred from the GSA Administrator to the DHS Secretary on February 28, 2003.[13] Within DHS, the chairmanship of the ISC was delegated to the Director of the Federal Protective Service (FPS) in January 2004.[14]

A 2004 report issued by GAO recommended that DHS direct the ISC to develop a plan that "identifies resource needs, implementation goals, and time frames for meeting the ISC's ongoing and yet-unfulfilled responsibilities."[15] GAO reported that standard operating procedures had been approved by agency members, and included new requirements for attendance and participation at ISC meetings. To address these issues, DHS (through the ISC) is creating and maintaining a centralized security database of all existing federal facilities. Since its transfer to DHS, the ISC has either updated or established the following standards or best practices:

- *Use of Physical Security Performance Measures*, gives policy guidance on metrics and testing for physical security programs;

- *Facility Security Level Determinations*, defines the criteria and process used in determining the security level of a federal facility;
- *Physical Security Criteria for Federal Facilities*, establishes a baseline set of physical security measures to be applied to all federal facilities;
- *Design-Basis Threat Report*, creates a profile of the type, composition, and capabilities of potential threats to federal facilities; and
- *ISC Best Practices for Safe Mail Handling*, identifies best practices for mail room operations in federal agencies and assists security managers in implementing safe mail-handling practices.[16]

In addition to its duties to coordinate federal security efforts and develop security standards for the construction of new federal facilities, the ISC has been assigned responsibility for reviewing federal agencies' physical security plans. Homeland Security Presidential Directive 7, issued December 17, 2003, required federal agencies "to identify and prioritize United States critical infrastructure and key resources and to protect them from terrorist attacks," and it assigned implementation responsibilities to DHS.[17] In July 2004, the ISC was designated to oversee and review each agency's physical security plan pertaining to protection of the nation's infrastructure and key resources. According to GAO, the ISC's successful completion of these new responsibilities would represent "a major step" toward carrying out its existing duties pertaining to compliance and oversight.[18] The ISC, however, does not have the authority to enforce standards on other federal departments and agencies. It relies on other federal entities' willingness and abilities to implement security standards and best practices.

EXECUTIVE BRANCH FACILITY SECURITY

Numerous agencies have responsibility for federal building security entities. Among them are such law enforcement agencies as the Central Intelligence Agency's (CIA) Security Protective Service, the Department of Defense's (DOD) Pentagon Police Directorate, and the State Department's Diplomatic Security Service's uniformed law enforcement officers. These agencies and facilities are usually limited in scope and size, involving a single location or a limited number of buildings. However, the primary agency for protecting federal facilities is the Federal Protective Service (FPS), which is responsible for protecting almost half (48%) of all GSA owned or leased property.[19]

Federal Protective Service

FPS, now within DHS's National Protection and Programs Directorate (NPPD),[20] is responsible for the protection and security of federally owned and leased buildings and property and of federal personnel.[21] In general, FPS operations focus on security and law enforcement activities that reduce vulnerability to criminal and terrorist threats. FPS protection and security operations include all-hazards based risk assessments; emplacement of criminal and terrorist countermeasures, such as vehicle barriers and closed-circuit cameras; law enforcement response; assistance to federal agencies through Facility Security Committees; and emergency and safety education programs. FPS also assists other federal agencies with additional security, such as assisting the U.S. Secret Service at National Special Security Events (NSSE).[22] FPS is the lead Government Facilities Sector Agency for the National Infrastructure Protection Plan.[23] Currently, FPS employs approximately 1,225 law enforcement officers, investigators, and administrative personnel, and administers the services of approximately 15,000 contract security guards.[24] Currently, FPS employs approximately 1,225 law enforcement officers, investigators, and administrative personnel. The Senate-reported version of FY2012 appropriations for DHS recommended \$1.2 billion for FPS, and the House-reported version recommended \$1.3 billion.[25]

Historical Overview and Current FPS Authority

The responsibility to protect federal buildings was given to the Federal Works Agency in June 1948.[26] Specifically, Congress authorized the Federal Works Administrator to appoint uniformed guards as special policemen with responsibility for "the policing of public buildings and other areas under the jurisdiction of the Federal Works Agency."[27] The special policemen were given the same responsibility as sheriffs and constables on federal property to enforce the laws enacted for the protection of persons and property, and to prevent "breaches of peace, and suppress affrays or unlawful assemblies."[28]

On June 30, 1949, the Federal Works Agency was abolished, and all of its functions, including the protection of federal buildings, were transferred to GSA[29] In September 1961, Congress authorized the GSA Administrator to appoint non-uniformed special policemen to conduct investigations in order to protect property under the control of GSA; enforce federal law to protect persons and property; and make an arrest without a warrant for any offense committed upon federal property if a policeman had reason to believe the offense was a felony and the person to be arrested was guilty of the felony.[30]

The GSA Administrator formally established the Federal Protective Service (FPS) in January 1971 through GSA Administrative Order 5440.46. FPS, as an official GSA agency, continued to protect federal property and buildings with both uniformed and non-uniformed policemen.

FPS was transferred to the Department of Homeland Security, and placed within U.S.

Immigration and Customs Enforcement (ICE), with enactment of the Homeland Security Act of 2002 (P.L. 107-296). The act required the DHS Secretary to "protect the buildings, grounds, and property that are owned, occupied, or secured by the Federal Government (including any agency, instrumentality, or wholly owned or mixed ownership corporation thereof) and persons on the property."[31] With the passage of FY2010 DHS appropriations, Congress authorized the transfer of FPS from ICE to the National Protection and Programs Directorate.[32] On October 29, 2009, DHS Secretary Janet Napolitano announced this transfer.[33]

Under current statutory provisions, FPS officers are authorized to

- enforce federal laws and regulations to protect persons and federal property;
- carry firearms;
- make arrests without a warrant for any offense against the United States committed in the presence of an officer or for any federal felony;
- serve warrants and subpoenas issued under the authority of the United States;
- conduct investigations, on and off federal property, of offenses that may have been committed against federal property or persons on the property; and
- carry out other activities for the promotion of homeland security as the DHS Secretary may prescribe.[34]

Federal Protective Service Use of Contract Security Guards

FPS's contract security guard responsibilities include federal building access control, employee and visitor identification checks, security equipment monitoring, and roving patrols of the interior and exterior of federal property.[35] Within the National Capital Region (NCR), FPS contracts with 54 private security guard companies to provide approximately 5,700 guards to protect 125 federal facilities. FPS issues task orders to contract security guard services that detail the terms and conditions under which the contract security guard services are to be provided. Some of these task orders include the identification of buildings requiring protection, specific guard post locations, and the hours and days of the week each post is to be staffed; whether security guards are to be armed; and the number of guards at each post. FPS currently employs approximately 15,000 contract security guards across the nation, and, according to the DHS Inspector General (DHS IG), contract guard services "represent the single largest item in the FPS operating budget, with an estimated FY2006 budget of $487 million."[36]

FEDERAL COURT FACILITY SECURITY

The safe conduct of court proceedings and the security of judges, court personnel, and visitors in courtrooms, as well as the safety of judges off-site, continue to be a concern. The 2005 murders of family members of a federal judge in Chicago; the killings of a state judge, a court reporter, and a sheriff's deputy at an Atlanta courthouse; and the 2006 sniper shooting of a state judge in his Reno office all spurred efforts to improve judicial security.[37] Other threats against judges and court facilities have not stopped. For example, in September 2009, a plan to bomb the Paul Findley Federal Building and Courthouse in Springfield, IL, was uncovered and an arrest was made. On January 4, 2010, a gunman wounded a deputy U.S. marshal and killed a court security officer at the Lloyd D. George U.S. Courthouse and Federal Building in Las Vegas. Possible threats in the first week of 2010 included suspicious substances in letters sent to courthouses in Alabama. These recent incidents may result in review and increased oversight of judicial security at court facilities to ensure that adequate protective policies, procedures, and practices are in place. Additionally, increased security enhancements may be necessary for federal courthouses where trials of individuals charged with acts of terrorism are to be held.

Each of the three branches of the federal government plays a unique role in helping to ensure the safety of judges and the security of the federal courts. The role of Congress is to authorize programs that enhance security, appropriate funds, and provide oversight of judicial security. The Judicial

Conference's Committee on Judicial Security monitors the security of the judiciary (including the protection of court facilities and proceedings, judicial officers, and court staff at federal court facilities and other locations) and makes policy recommendations to the conference. The Administrative Office of the U.S. Courts implements Judicial Conference policies, including security matters.

By statute,[38] the United States Marshals Service within the Department of Justice has primary responsibility for the security of the federal judiciary, including the safe conduct of court proceedings, as well as the security of federal judges and court personnel at court facilities and off-site. USMS is charged with the protection and security of more than 2,000 federal judges and approximately 5,250 other court officials[39] at over 400 court facilities nationwide.[40] Within USMS, the Judicial Security Division (JSD) is specifically responsible for providing security services and staff support for the federal judiciary, including personal protection for judges and physical security for federal courthouses. Other space in the court facilities under the control of USMS includes holding cells adjacent to courtrooms, interview rooms used by attorneys and prisoners, cellblocks, prisoner elevators, and office space for USMS use. An appointed U.S. marshal, confirmed by the Senate, has security responsibility in each of the 94 federal judicial districts and the District of Columbia Superior Court. District U.S. marshals provide and oversee security of the judiciary using USMS resources and court security officers (CSO), who are employees of private security companies under contract with USMS. Over 4,500 CSOs provide various types of security (e.g., fixed posts, roving patrols, entry screening, and mail and package screening) in courthouses and at multi-tenant facilities. Also under USMS jurisdiction are the design, installation, and maintenance of security systems, and the oversight of communications equipment.

USMS conducts investigations of threats made against federal judges, U.S. attorneys, court staff, and their family members to determine the level of security that is necessary for developing security plans. In accordance with these findings, USMS assigns the required resources to ensure the safety of these people. A deputy marshal is required to attend any session of court at the request of the presiding judge.[41] A judicial security inspector (a senior-level deputy marshal) is assigned to each judicial district to evaluate courthouse security and procedures and to coordinate scheduling, posting, and other matters related to CSOs. The inspectors also conduct security surveys at judges' homes and recommend improvements. On June 1, 2004, USMS established the Office of Protective Intelligence (OPI) to review and analyze intelligence information about the security of those under USMS protection. OPI issues daily security advisories, intelligence bulletins, and law enforcement alerts to USMS district offices and senior staff at headquarters so that protective measures can be taken. When threats are made, USMS works with the Federal Bureau of Investigation (FBI) to evaluate the threats.

Within the Department of Homeland Security, FPS has overall responsibility for security in GSA-managed, multi-tenant federal buildings. When the buildings include court facilities, USMS and FPS share security responsibilities; this is authorized by a series of memoranda of agreement and understanding (MOA and MOU) between GSA and DOJ.[42] When the court is the sole tenant in a GSA-managed building, USMS has primary responsibility for security, although FPS may provide some support for the perimeter security, or it may delegate this responsibility to USMS. The manner in which the responsibilities are shared varies case by case, depending on the differing requirements of tenants, functions, and locations of occupied space. These shared responsibilities and jurisdictions at individual court-occupied buildings are further determined by agreements (sometimes in writing), and coordinated to avoid duplication. Generally, USMS is responsible for and controls access to judicial space, while FPS is primarily responsible for perimeter security and for other interior space that is not court-related space. FPS conducts risk assessments of multi-tenant buildings to deter threats and take countermeasures. Uniformed FPS officers and hired contract guards (similar to court security officers) protect the buildings and their assets, and investigate crime at the facilities. Other than perimeter responsibilities, FPS duties may include visitor entry processing, roving patrols, garage access control, and mail and package screening.[43]

These principal entities communicate and coordinate at the national and district levels to ensure the security of the courts. At the national level, the Judicial Conference's Committee on Judicial Security coordinates security issues involving the federal courts with USMS, DOJ, and DHS. According to USMS, the Marshals Service works daily with the AOUSC Office of Court Security and the Office of Facilities and Security, and the Committee on Judicial Security also consults and coordinates over national and district-level security matters. At semi-annual meetings, the Committee on Judicial Security and USMS senior management discuss security, legal, and budget issues. In addition, USMS and AOUSC hold working sessions to discuss issues that include the purchase and installation of security systems, CSO staffing, and budget matters. At the local level, U.S. marshals routinely meet with the district chief judge at court security committee meetings including representatives from the magistrate, district, and bankruptcy courts (and sometimes circuit judges and U.S. attorneys) to review and implement security plans. AOUSC and USMS also consult on security considerations (e.g., design and installation of security systems) in the construction of new or renovated courthouses.

On January 5, 2009, USMS implemented a pilot program to assume primary responsibility for perimeter security at selected courthouses that were previously the responsibility of the FPS. This pilot was undertaken in accordance with FY2009 enacted legislation[44] as a result of the judiciary's concerns that FPS was providing inadequate perimeter security. The pilot program, expected to cover an 18–month period, includes five courthouses located in Chicago, Detroit, Phoenix, New York, and Tucson, and two in Baton Rouge. The judiciary submitted a report to the House and Senate Appropriations Subcommittees on Financial Services and General Government on the implementation progress of the pilot program and is currently working with USMS on assessment tools for the program.

Supreme Court

As the Supreme Court's general manager, paymaster, and chief security officer, the Marshal of the Supreme Court oversees the administration and operations of the Court building.[45] The Marshal manages over 200 Court employees and supervises the federal property used by the Court. The Marshal also directs the Supreme Court Police Force, which comprises a chief of police and approximately 80 officers. The police force jurisdiction covers the Court building, its grounds, and adjacent streets.

Coordination of Federal Building Security

Federal building security includes such activities as the daily interaction of FPS and its federal customers, the coordination between USMS and the FPS in federal multi-use buildings, and the federal agency interaction with contract security guard companies. It should be noted that some Members of Congress have state and district offices in multi-tenant federal buildings and recently the FBI has coordinated with Members due to an increase in threats to Members and their offices.[46] Federal agencies communicate with one another and state, local, and private sector entities to coordinate federal building security. It is important to note that the federal government's communication of potential and imminent terrorist and criminal threats to states, localities, and private sector entities is an important aspect of federal building security because the majority of federal agencies and departments lease, build, and occupy facilities located in local jurisdictions and are not segregated from the general populace, private industries and businesses, and state and local government facilities. Not only would local jurisdictions be susceptible to collateral damage in a terrorist attack on a federal building, but some federal agencies and departments also rely on state and local law enforcement entities in the event of criminal or terrorism activities at federal facilities.[47]

One established way the federal government communicated threats was through the use of the Homeland Security Advisory System (HSAS), which was managed by DHS. HSAS, established on March 12, 2002, was a color-coded terrorist threat warning system. The system, which federal departments and agencies were required to implement and use, provided recommended protective measures for federal departments and agencies to prevent, prepare for, mitigate against, and respond to terrorist attacks. DHS disseminated HSAS terrorist threat warnings to federal departments, state and local agencies, the public, and private-sector entities. DHS, however, only provided protective measures for federal departments. This dissemination of warnings was conducted through multiple communication systems and public announcements. HSAS had five threat levels: low, guarded, elevated, high, and severe.[48] In 2009, DHS's Homeland Security Advisory Council established a task force to review the HSAS and recommend changes to the administration and use of the system.[49] Upon review, DHS replaced the HSAS with the National Terrorism Advisory System (NTAS). NTAS is a new system that is to communicate terrorism threat information by providing "timely, detailed information to the public, government agencies, first responders, airports and other transportation hubs, and the private sector."[50]

Some federal entities, in response to targeted and specific threats, have developed mechanisms for notifying other federal departments and agencies, such as the U.S. Nuclear Regulatory Commission's Office of Nuclear Security and Incident Response, which coordinates with DHS, the federal intelligence and law enforcement communities, and the Department of Energy (DOE). In 2005, John E. Lewis, Deputy Assistant Director of the FBI's Counterterrorism Division, testified before the House Committee on Homeland Security about the FBI's coordination with other federal agencies concerning potential nuclear threats or incidents. Mr. Lewis stated that the FBI has developed liaison relationships with DHS, DOE, and DOD, and he detailed how the FBI and these departments would coordinate their response efforts if there was a nuclear threat or incident.[51]

Within DHS, the Office of Operations Coordination is responsible for monitoring the nation's security situation daily, through the National Operations Center (NOC), and coordinating activities among DHS, governors, homeland security advisors, law enforcement entities, and critical infrastructure operators. Information on domestic incident management is shared with Emergency Operations Centers at federal, state, and local levels through the Homeland Security Information Network (HSIN), and state and local intelligence fusion centers.[52]

FEDERAL BUILDING SECURITY ISSUES

Due to recent attacks on federal buildings and continued terrorism threats, Congress may wish to address issues associated with federal building security. Some of these issues include FPS's operations and use of contract security guards, coordination and sharing of federal building security information, Facility Security Committees, and appropriations and resources. These issues are discussed below.

Federal Protective Service Operations and Use of Contract Security Guards

The threat of terrorism since the September 11, 2001, attacks has increased emphasis on the physical security of federal property and congressional interest in FPS.[53] Since 2009, GAO has issued three reports about FPS, one on FPS's use of contract security guards[54] and two on FPS's operations to address federal facility vulnerabilities.[55]

Federal Protective Service Operations

In November 2009, GAO identified the following concerns about FPS's operations:

- FPS does not have a risk management framework that couples threats and vulnerabilities with resource requirements;

- FPS lacks a strategic human capital plan to guide its current and future workforce planning efforts;[56]
- FPS lacks a systematic approach for using technology to reduce risk to federal facilities;
- FPS is inconsistent in sharing information and coordinating security with GSA and tenant agencies; and
- FPS lacks a reliable data management system for accurately tracking performance measurement and testing.[57]

One GAO recommendation is for FPS to improve its use of a fee-based system by developing an accurate method of accounting for the cost of providing security services to tenant agencies and ensuring that its fee structure takes into consideration the varying levels of risk and service provided at GSA facilities.[58] Congress may wish to address the implementation of this recommendation by requiring FPS specifically, and DHS generally, to develop an accurate method for assessing security service costs through statutory or conference language.

Additionally, GAO recommended an evaluation of whether continued use of the current fee-based system or another funding mechanism would be the most appropriate method for funding FPS operations.[59] Congress might want to require the DHS IG to review the use of the fee-based system versus the method of directly providing appropriations to FPS. Alternatively, Congress could determine without further review to begin a direct appropriation for FPS operations through statutory language in annual DHS appropriations. This approach would possibly reduce the amount of appropriations GSA and tenant agencies currently receive to pay FPS for security operations.

Considering all of these issues, Congress may want to review FPS operations further through oversight hearings or require FPS and DHS to report on what actions, if any, the agency is taking to address GAO and DHS IG findings. Further review and hearings may not, however, immediately ameliorate continuing FPS shortcomings.

Concerns About Federal Protective Service Use of Contract Security Guards

GAO identified concerns with FPS's use of contract security guards, including that

- FPS does not fully ensure that its contract security guards have the training and certifications required to secure federal facilities;

- FPS does not have a completely reliable system for monitoring and verifying contract guard training and certification requirements;
- FPS does not have specific national guidance on when and how contract guard inspections should be performed; and
- FPS inspections of contract security guard posts at federal facilities are inconsistent, and the quality of the inspections varies across FPS regions.[60]

FPS has implemented some actions in response to GAO's findings. According to FPS officials, these actions include authorizing overtime to monitor contract security guards during non-routine business hours and requiring penetration tests to identify weaknesses at access control contract security guard posts. Additionally, FPS has implemented a new directive developed to clarify FPS responsibilities for conducting and reporting the results of inspections and evaluations.[61]

In FY2010, Congress attempted to address FPS's use of contract security guards by requiring FPS to maintain no fewer than 1,200 full-time equivalent staff and 900 full-time police officers, investigators, inspectors, area commanders, and special agents.[62] This requirement could increase FPS's oversight of contract security guards; however, increasing the number of FPS law enforcement officers may not solve problems immediately because of the time required to inspect contract security guard operations, to identify continuing shortcomings, and to train FPS and contract security guard personnel. Additionally, increasing the number of FPS law enforcement personnel could further strain FPS resources by increasing the amount of personnel benefits afforded to federal employees.

Coordination and Sharing of Federal Building Security Information

Terrorism threat information sharing and coordination of federal, state, and local government security operations are multi-faceted endeavors that require constant attention and are immediately reviewed, and possibly revised, following an attempted or successful terrorist attack, such as the recent attempted bombing of an airplane on December 25, 2009.[63] Federal facilities and agencies sharing terrorism threat information and coordinating facility security are specific and integral parts of this government endeavor to ensure the nation's security. Congressional action on terrorism information sharing includes passage of the Intelligence Reform and Terrorism Prevention Act of 2004, which mandated the creation of an Information Sharing Environment (ISE).[64] The ISE is to facilitate the sharing of terrorism information among federal, state, local, and private sector entities through the use of policy guidelines and technologies.[65] However, problems have arisen related to the coordination and sharing of federal building security information.

GAO, in November 2009 testimony before the House Committee on Homeland Security, stated that even though FPS and GSA management officials have established communication processes, information sharing at the regional and facility levels is inconsistent, and FPS and GSA disagree overall about what information should be shared. As an example, GAO cited a memorandum of agreement between DHS and GSA that specified that FPS will provide quarterly briefings at the regional level; however, this has not been done consistently across all FPS regions. GSA security officials stated that the briefings that did occur primarily focused on crime statistics and did not constitute comprehensive threat analyses.[66]

Additionally, on September 30, 2009, DHS Secretary Janet Napolitano stated, before the Senate Homeland Security and Governmental Affairs Committee, that there is no single process or system for federal, state, and local entities to receive or share terrorism intelligence and threat information. The Secretary stated that the present system of sharing information is not streamlined, that it is a "work in progress," and that this may be the result of the security classification of the information. Also, the Secretary said that state and local officials may be confused about where to obtain terrorism threat information.

To address the issue of coordination and sharing of terrorism threat information among federal facilities, Congress could choose to require, through statutory language, that federal agencies report periodically on this matter to the committees of jurisdiction. Additionally, Congress could request that GAO revisit the ISE and its implementation since 2004. In 2008, GAO reported limited success with the implementation and noted that the ISE lacked guidance on ensuring accountability and assessing progress.[67] Additionally, Congress could request information on how federal agencies train personnel (specifically, building security managers and officials) on the sharing of terrorism threat information.

Congress could also require a review of the coordination and integration of information sharing systems used by federal agencies, such as Law Enforcement Online and the Homeland Security Information Network. This review might identify what systems are utilized most and what systems appear effective. The review might disclose which federal agencies are involved, and at what level, in each system. On this point, GAO noted in 2007 that when "identical or similar types of information are collected by or submitted to multiple agencies, integrating or sharing this information can lead to redundancies."[68]

None of these options, however, address the issue of how specific federal facilities interact with federal security entities such as FPS, or what type of informal coordination is conducted daily to ensure the safety and well-being of federal employees, and members of the general public who visit federal facilities.

Facility Security Committees

When a federal court is in a multi-tenant building, the court's representative is a member of the Facility Security Committee (FSC) (previously known as the Building Security Committee). Each FSC, made up of tenants in the building, considers and makes decisions on building security matters.

FSCs were mentioned at the November 18, 2009, House Committee on Homeland Security hearing on Federal Protective Service Transition.[69] The committee chair raised the issue of the potentially dangerous items that are allowed to be brought into federal buildings. In July 2009, GAO staff reported that they were able to smuggle bomb-making components into 10 high-security federal facilities in four different cities and successfully assemble the items inside the building. The components were not on the prohibited list for those facilities. Reportedly, FSCs in each federal building determine what items are prohibited, and there is no standard list of prohibited items. FPS makes security assessments and presents them to FSCs, but, according to the FPS director, FPS does not determine the prohibited items list.[70]

Congress might consider whether the current process for determining which items should be prohibited is sufficient, and whether the input provided by various tenants, including the judiciary—as well as recommendations by USMS, FPS, GSA, and others—is effective. Further, consideration might be given to mandating a standard list of prohibited items to enhance security at federal facilities and to establishing regular evaluations of the list.

Research[71] also indicated that few across-the-board standards have been established for the FSCs. The Interagency Security Committee formed a working group to examine and issue a document to address FSC operations.

While each federal facility may have different or unique security administrative challenges, the lack of standards in FSC administrative operations could imperil security. Among the areas that could be examined are the following:

- FSC membership composition, and the designation and authority of the committee chair;
- voting issues, including how a quorum and majority vote are determined;
- whether one vote for each tenant is fair or whether votes should be proportional to tenant space;
- whether the formula for allocating how much each tenant pays for security enhancements, now generally based on square footage, is equitable or appropriate;
- whether a minimum number of mandatory regularly scheduled meetings of the FSC (and others on an as-needed basis) should be required;
- whether there is a need for possible standards for written records of FSC meetings and other recordkeeping requirements;
- whether there is a system for tenants to appeal decisions and resolve possible disagreements;
- whether each tenant has efficient processes for securing the approval of its headquarters office with regard to security enhancement requests so that FSC can act to implement improvements without delay from one or more tenants;
- whether FSCs have adequate and timely communication with federal, state, and local law enforcement organizations; and
- whether there should be regular congressional or ISC review of FCS operations and reporting requirements.

Legislation in 112th Congress: Federal Protective Service

Early in the 112th Congress, legislation was introduced in both the House and Senate to improve the ability of FPS to protect federal facilities, the employees who work there, and public visitors. Both bills are similar to legislation introduced in the previous Congress but no further action was taken before the 111th Congress adjourned.[72]

House Bill (H.R. 176)

On January 5, 2011, Representative Bennie G. Thompson introduced H.R. 176, the Federal Protective Service Improvement and Accountability Act of 2011. On January 31, 2011, the bill was referred to the House Transportation and Infrastructure Committee; House Homeland Security Committee; and subsequently referred to the Subcommittee on Emerging Threats, Cybersecurity, and Science and Technology, where it is pending.

H.R. 176 would reform the FPS workforce, security practices, and contract oversight capacity. Following is a summary of the bill's main provisions:

- The DHS Secretary would be directed to maintain no fewer than 1,350 full-time equivalent positions in the FPS inspector force (fully trained federal law enforcement officers). These positions would be classified as Federal Facility Security Officers (responsible for security assessment), and law enforcement officers (responsible for physical law enforcement and investigations).
- The DHS Secretary would establish a FPS contract oversight force to monitor contracts, contractors, and contract guards, and require minimum training and certification standards for security guard services at FPS-protected facilities.
- The FPS Director would begin a one-year pilot program (within six months after the enactment of the act) to research the conversion of contract guard positions to federal employee positions at the highest-risk federal facilities.
- The GAO Comptroller General would review periodically the pilot program and the performance of federal facility security guards in the program, and submit a final report of its evaluation to Congress.
- The GAO Comptroller General would review the current FPS fee-based funding system, and recommend changes or alternatives, as appropriate, to the system.

H.R. 176 also expresses the sense of Congress that specified security standards for federal facilities established by the Interagency Security Committee should be implemented for all federal facilities.

Senate Bill (S. 772)

On April 8, 2011, Senator Joseph I. Lieberman introduced (for himself and Senators Susan M. Collins and Daniel K. Akaka) S. 772, the Supporting

Employee Competency and Updating Readiness Enhancements for Facilities Act of 2011 (SECURE Facilities Act of 2011). The purpose of the bill is to protect federal employees and public visitors by improving the security of federal facilities and modernizing FPS. The bill was referred to the Senate Homeland Security and Governmental Affairs Committee on the same day. On May 11, 2011, the committee adopted by a vote of 10-5, an amendment Senator Tom Coburn offered to require the DHS Secretary to offset each new full-time FPS employee position with the reduction of one full-time DHS employee position.[73] On May 18, 2011, the Senate committee reported the bill favorably, as amended, by voice vote.

S. 772 would increase the guard workforce, require training and regular assessments of guard capabilities, and stabilize the management of FPS. Following is a summary of the bill's main provisions:

- The DHS Secretary would ensure that FPS maintain at any time no fewer than 1,371 full-time equivalent employees (including no fewer than 950 in-service field staff) in FY2012. The DHS Secretary is also to ensure that there would be no fewer than 1,200 full-time employees at any time (including 900 in-service field staff). After FY2012, the DHS secretary would submit a report to the appropriate congressional committees if there is a decrease in the number of full-time equivalent employees and provide a revised model projected for future fiscal years on the number of full-time equivalent employees.
- FPS would increase the training provided to contract guards, maintain testing programs to assess training and the security of federal facilities, and establish procedures for retraining or terminating guards.
- FPS officers would be authorized to carry firearms on or off duty.
- The DHS Secretary would coordinate with the ISC, to develop standards for checkpoint detection technologies for explosives and other threats at FPS-protected federal facilities.
- ISC's authority to enforce compliance on security standards for the federal buildings would be increased. The DHS Secretary would submit a report to the appropriate congressional committees if any facility is determined to be in noncompliance with the ISC security standards.
- Federal agencies' representatives serving on a Federal Security Committee would be trained on security matters, in accordance with standards established by ISC. Other requirements for each committee

include meeting on a quarterly basis, or more frequently (as determined by the committee chair), and maintaining official records of the meetings.

- The DHS Secretary would be required to submit a report to the appropriate congressional committees within 180 days after enactment of the act on the FPS funding system, including recommendations for alternatives, including direct appropriations, or a combination of fee collections, security charges, and appropriations.

Appropriations and Resources

If the Administration decided to hold trials of individuals[74] charged with terrorist acts in federal civilian court rather than in military tribunals, enhancing communication and coordination of all law enforcement agencies (federal, state, and local) and the judiciary would be critical. The planning and implementation of additional security enhancements, including staff, training, technology, and equipment, would be necessary for the security and safety of all parties involved. A systematic method for identifying and addressing concerns of the local community and officials might include assessing the disruptive impact security measures might have on the businesses and residences in the area, and on traffic around and leading to the courthouses. Depending on the location of such trials, airspace surveillance and extension of the perimeter for security might be considered.

Congress might take into account the unique jurisdictional responsibilities of each entity, including national intelligence agencies. Establishing a system for entities to coordinate expenditures in a timely manner might be necessary because such trials could continue for several years. Planning for multiple years of appropriations might be needed to enable the entities to continue to fulfill their functions.

Congressional oversight could be critical to ensure that funds and resources are maximized under fiscal constraints. If sufficient funding is not provided in a timely manner to the federal entities involved, Congress might consider whether authorities are in place for entities to transfer resources currently devoted to other programs, and whether such transfers might adversely affect the performance of other missions.

Congress might also consider whether funding FPS directly could provide it with more stable and predictable funding than its current reliance on fees from other agencies and the judiciary. In addition, direct appropriations to FPS might reduce administrative costs for both FPS and the federal courts.

CONCLUSION

The federal government faces, daily, the task of securing a portfolio comprising 446,000 buildings. Accordingly, Congress could address concerns, some of which are addressed in this chapter, to ensure effective federal agency operations and the health, well-being, and safety of federal employees and the public.

End Notes

[1] U.S. Department of Justice, U.S. Marshals Service, *Vulnerability Assessment of Federal Facilities*, Washington, DC, June 28, 1995, p. 1-1.

[2] U.S. Government Accountability Office, *Building Security: Interagency Security Committee Has Had Limited Success in Fulfilling Its Responsibilities*, GAO-02-1004, September 2002, p. 5.

[3] U.S. Department of Justice, U.S. Marshals Service, *Vulnerability Assessment of Federal Facilities*, Washington, DC, June 28, 1995.

[4] U.S. President (Clinton), "Memorandum on Upgrading Security at Federal Facilities," *Public Papers of the Presidents of the United States*, vol. I, June 28, 1995, pp. 964-965.

[5] Both bills are similar to legislation introduced in September 2010, but no further action was taken before the 111th Congress adjourned. These bills were H.R. 6122, the Federal Protective Service Improvement and Accountability Act of 2010, introduced by Representative Bennie G. Thompson (for himself, and Representatives Sheila Jackson Lee and Laura Richardson) on September 14, 2010. It was referred to the House Committees on Transportation and Infrastructure and Homeland Security. On September 17, 2010, H.R. 6122 was referred to the House Subcommittee on Transportation Security and Infrastructure Protection. On September 20, 2010, Senator Joseph I. Lieberman (for himself and Senators Susan M. Collins, Daniel K. Akaka, and George V. Voinovich) introduced S. 3806, the Supporting Employee Competency and Updating Readiness Enhancements for Facilities Act of 2010. It was referred to the Senate Homeland Security and Governmental Affairs Committee. Although S. 3806 was placed on the Senate calendar, no further action was taken. In addition, on April 15, 2010, Representative Charles W. Dent introduced (for himself and Representatives Steve Austria, Peter T. King, Michael T. McCaul, and Pete Olson). H.R. 5053, the Federal Protective Service Reform and Enhancement Act of 2010. The purpose of H.R. 5053 was to improve security for the prevention of terrorist activities and for the promotion of homeland security, and for other purposes. On the same day, H.R. 5053 was referred to the House Transportation and Infrastructure Committee, and subsequently referred to the

Subcommittee on Economic Development, Public Buildings and Emergency Management. No further action was taken on these three bills before the 111th Congress adjourned.

[6] U.S. Department of Justice, U.S. Marshals Service, *Vulnerability Assessment of Federal Facilities*, Washington, DC, June 28, 1995.

[7] Executive Order 12977, "Interagency Security Committee," 60 *Federal Register* 54411-54412, October 24, 1995.

[8] Ibid., p. 54412.

[9] This document is available only from the ISC. Contact information for the ISC is at http://www.dhs.gov/files/ committees/gc_1194977813020.shtm.

[10] U.S. General Services Administration, Public Building Service, Memorandum for Assistant Regional Administrators, *Implementation of the Interagency Security Committee (ISC) Design Criteria Regarding Site Selection*, Washington, DC, April 26, 2002, pp. 1-2.

[11] U.S. Government Accountability Office, *Building Security*, GAO-02-1004, 2002, pp. 7-11.

[12] 116 Stat. 2135.

[13] Executive Order 13286, "Amendment of Executive Orders, and Other Actions, in Connection with the Transfer of Certain Functions to the Secretary of Homeland Security," 68 *Federal Register* 10624, March 5, 2003.

[14] U.S. Government Accountability Office, *Homeland Security: Further Actions Needed to Coordinate Federal Agencies' Facility Protection Efforts and Promote Key Practices*, GAO-05-49, September 6, 2004, p. 9.

[15] Ibid., pp. 47-48.

[16] These documents are "For Official Use Only" and can be obtained from the ISC. Summaries of these documents can be found at http://www.dhs.gov/files/committees/gc_1194978 268031.shtm.

[17] HSPD-7, "Directive on Critical Infrastructure Identification, Prioritization, and Protection," *Weekly Compilation of Presidential Documents,* vol. 39, December 17, 2003, p. 1816.

[18] U.S. Government Accountability Office, *Homeland Security: Further Actions Needed to Coordinate Federal Agencies' Facility Protection Efforts and Promote Key Practices*, GAO-05-49, September 6, 2004, p. 11.

[19] U.S. Government Accountability Office, *Homeland Security: Federal Protective Service Should Improve Human Capital Planning and Better Communicate with Tenants*, GAO-09-749, July 2009, p. 22. Other GSA-controlled facilities use an internal office within the tenants' agency (29%). The remainder rely on specific entities, such as GSA's Building Security and Policy Division, the U.S. Marshals Service, U.S. Postal Service Inspection Services, and private contractors (23%).

[20] FPS was transferred to NPPD from Immigration and Customs Enforcement following the enactment of FY2010 DHS appropriations (P.L. 111-83). 123 Stat. 2157.

[21] 40 U.S.C. 1315.

[22] For information on NSSEs, see CRS Report RS22754, *National Special Security Events*, by Shawn Reese.

[23] Information on the NIPP is at http://www.dhs.gov/xprevprot/programs/editorial_0827.shtm.

[24] U.S. Department of Homeland Security, National Protection and Programs Directorate, *Federal Protective Service: Fiscal Year 2011 Congressional Justification*, Washington, DC, February 2010, pp. FPS-1.

[25] H.R. 2017.

[26] 62 Stat. 281.

[27] Ibid.

[28] Ibid.

[29] 63 Stat. 380.

[30] P.L. 87-275, Sec. 5, 40 U.S.C. §318.

[31] 40 U.S.C. § 1315(a). The DHS Secretary was authorized to designate DHS employees, including those transferred from FPS, as officers with responsibility for protecting federal property. Some federal buildings, however, are protected by guards who are not part of FPS, such as the buildings of the U.S. State Department, which has its own uniformed law enforcement officers.

[32] P.L. 111-83, Title III. 123 Stat. 2157.

[33] U.S. Department of Homeland Security, "Secretary Napolitano Announces Transfer of Federal Protective Service to National Protection and Programs Directorate," press release, October 29, 2009, http://www.dhs.gov/ynews/releases/ pr_1256829032272.shtm.

[34] 40 U.S.C. § 1315(b)(2).

[35] U.S. Department of Homeland Security, Office of Inspector General, *Federal Protective Service Needs to Improve Its Oversight of the Contract Guard Program*, OIG-07-05, October 2006, p. 2.

[36] Ibid.

[37] In the 110th Congress, the President signed into law the Court Security Improvement Act of 2007 (P.L. 110-177), which was designed to enhance security for judges and court personnel as well as courtroom safety for the public.

[38] 28 U.S.C. § 566(a).

[39] USMS also provides protective details for judges and others who are targets of threats and attacks, and provides other law enforcement services for DOJ. For example, USMS is responsible for (1) providing protection for witnesses who testify for the government in cases involving organized crime and other significant criminal activity, (2) transporting criminal defendants to and from court appearances, (3) arranging for space in detention facilities to house pre-sentenced criminals, and (4) managing and disposing of forfeited properties acquired by criminals through illegal activities. For more information, see http://www.usmarshals.gov/duties.

[40] U.S. Department of Justice, United States Marshals Service, "Fact Sheet: Facts and Figures," USMS Pub. No. 21-H, revised February 9, 2009, Washington, DC, p. 1.

[41] As federal law enforcement officers, deputy marshals have other responsibilities, including criminal investigations, fugitive apprehension, witness protection, prisoner transportation, and execution of court orders.

[42] The December 1997 "Memorandum of Agreement for Court Security Between the GSA, USMS and AOUSC" defined each agency's area of responsibility for judicial security. According to the MOA, the three parties recognized that a cooperative effort was needed to provide the federal courts with the necessary security. This MOA stated, "The requirements of this joint effort are delineated in the March 1982, 'Report of the Attorney General's Task Force on Security'; the March 1982, 'Joint Statement of the Chief Justice and the Attorney General before the Judicial Conference of the United States'; the December 1982, Delegation of Procurement Authority from GSA to the Department of Justice (DOJ); the June 1995 DOJ report entitled 'Vulnerability Assessment of Federal Facilities'; and the 1997 Delegation of Authority from GSA to DOJ delegating USMS authority to determine and provide the appropriate level of perimeter access control at all GSA-controlled facilities that house a judicial officer." When FPS was transferred from GSA to DHS in 2003, the MOA was reaffirmed by a Memorandum of Understanding (MOU), which stated that the terms of the 1997 MOA would continue without interruption with DHS assuming the responsibilities transferred from GSA. Parties to the MOU were DOJ, DHS, and AOUSC

(signed by then-Attorney General John Ashcroft on November 20, 2003; then-director of AOUSC Leonidas R. Mecham on November 21, 2003; and then-Secretary of DHS Tom Ridge on January 21, 2004).

[43] Among FPS protective and security capabilities are (1) specialized response capabilities (e.g., canine, hazardous materials, and weapons of mass destruction response teams); (2) intelligence-sharing and investigative collaboration with law enforcement agencies at local, state, and federal levels; (3) key participation in federal anti-terrorism task forces; and (4) continuous monitoring of facility alarms and emergencies through FPS remote dispatch control centers.

[44] Omnibus Appropriations Act, 2009 (P.L. 111-8), Sec. 306, Title III, 123 Stat. 648.

[45] The Court building is U.S. government property, which was completed in 1935 as a permanent home to the Supreme Court.

[46] Emily Yehle, "FBI Reaches Out to Senator's State Offices," *Roll Call*, April 2, 2010.

[47] U.S. Government Accountability Office, *Homeland Security: Federal Protective Service Should Improve Human Capital Planning and Better Communicate with Tenants*, GAO-09-749, July 2009, pp. 5-6.

[48] Information on HSAS is available at http://www.dhs.gov/files/programs/Copy_of_press_release_0046.shtm.

[49] The task force's report and recommendations are available at http://www.dhs.gov/xlibrary/assets/hsac_task_force_report_09.pdf.

[50] http://www.dhs.gov/files/programs/ntas.shtm. For further information on NTAS, see http://www.dhs.gov/files/ publications/ntas-public-guide.shtm.

[51] Testimony of John E. Lewis, FBI Deputy Assistant Director, Counterterrorism Division, in U.S. Congress, House Committee on Homeland Security, Subcommittee on Prevention of Nuclear and Biological Attack, *Nuclear Incident Response Teams*, 109th Cong., 1st sess., October 27, 2005, Serial No. 109-50 (Washington: GPO, 2007).

[52] U.S. Department of Homeland Security, "Office of Operations Coordination," at http://www.dhs.gov/xabout/ structure/editorial_0797.shtm. For further information on homeland security information sharing, see CRS Report RL34070, *Fusion Centers: Issues and Options for Congress*, by John Rollins.

[53] U.S. Government Accountability Office, *Federal Real Property: An Update on High Risk Issues*, GAO-09-801T, July 15, 2009, p. 14.

[54] U.S. Government Accountability Office, *Homeland Security: Preliminary Results Show Federal Protective Service's Ability to Protect Federal Facilities Is Hampered by Weaknesses in Its Contract Security Guard Program*, GAO-09- 859T, July 8, 2008.

[55] U.S. Government Accountability Office, *Homeland Security: Federal Protective Service Has Taken Some Initial Steps to Address Its Challenges, but Vulnerabilities Still Exist*, GAO-09-1047T, September 23, 2009; and *Homeland Security: Greater Attention to Key Practices Would Help Address Security Vulnerabilities at Federal Buildings*, GAO-10-236T, November 18, 2009.

[56] U.S. Government Accountability Office, *Homeland Security: Federal Protective Service Has Taken Some Initial Steps to Address Its Challenges, but Vulnerabilities Still Exist*, GAO-09-1047T, September 23, 2009, p. 1.

[57] U.S. Government Accountability Office, *Homeland Security: Greater Attention to Key Practices Would Help Address Security Vulnerabilities at Federal Buildings*, GAO-10-236T, November 19, 2009, p. 1.

[58] Ibid., p. 6.

[59] Ibid.

[60] U.S. Government Accountability Office, *Homeland Security: Preliminary Results Show Federal Protective Service's Ability to Protect Federal Facilities Is Hampered by Weaknesses in Its Contract Security Guard Program*, GAO-09- 859T, July 8, 2008, p. 1.

[61] Ibid., pp. 16-17.

[62] P.L. 111-83, Title III, 123 Stat. 2157.

[63] For more information on terrorism information sharing, see CRS Report R40901, *Terrorism Information Sharing and the Nationwide Suspicious Activity Report Initiative: Background and Issues for Congress*, by Jerome P. Bjelopera; and CRS Report RL33873, *Sharing Law Enforcement and Intelligence Information: The Congressional Role*, by Richard A. Best Jr.

[64] P.L. 108-458, Sec. 1016(b), 118 Stat. 3665.

[65] Ibid. For more information on ISE, see CRS Report R40901, *Terrorism Information Sharing and the Nationwide Suspicious Activity Report Initiative: Background and Issues for Congress*, by Jerome P. Bjelopera.

[66] U.S. Government Accountability Office, *Homeland Security: Greater Attention to Key Practices Would Help Address Security Vulnerabilities at Federal Buildings*, GAO-10-236T, November 18, 2009, pp. 13-14.

[67] U.S. Government Accountability Office, *Information Sharing: Definition of the Results to Be Achieved in Terrorism-Related Information Sharing Is Needed to Guide Implementation and Assess Progress*, GAO-08-637T, July 23, 2008, p. 1.

[68] U.S. Government Accountability Office, *Homeland Security: Federal Efforts Are Helping to Alleviate Some Challenges Encountered by State and Local Information Fusion Centers*, GAO-08-35, October 30, 2007, p. 1.

[69] U.S. Congress, House Committee on Homeland Security, *Federal Protective Service Transition,* hearing, 111th Congress, 1st sess., November 18, 2009. Based on the *Congressional Quarterly* transcript of the hearing, accessible by subscription only at http://www.cq.com/ display.do?dockey=/cqonline/prod/data/docs/html/transcripts/congressional/ 111/congressional transcripts111-00003252978.html@committees&metapub=CQ-CONGTRANSCRIPTS& searchIndex=1&seqNum=6.

[70] Ibid.

[71] Research included various government documents, and discussions with the Administrative Office of the U.S. Courts (January 26 and 28, 2010), GSA (February 17, 2010), and other organizations.

[72] These bills were H.R. 6122, Federal Protective Service Improvement and Accountability Act of 2010, S. 3806, the Supporting Employee Competency and Updating Readiness Enhancements for Facilities Act of 2010, and H.R. 5053, the Federal Protective Service Reform and Enhancement Act of 2010.

[73] The committee voted against adopting two other amendments. By a vote of 7-7 (a tie vote fails), it defeated Senator John McCain's amendment to prohibit awarding a sole-source contracts to an Indian tribe, Alaska Native-owned, and Native Hawaiian-owned corporation (that exceeds $5 million for the procurement of services, and $6.5 million for the procurement of property, unless the DHS Secretary submits a written justification to Congress). The committee, by voice vote, did not approve Senator Coburn's amendment to eliminate the requirement for a feasibility report on federalizing the FPS contract guard workforce.

[74] Such individuals may include the September 11, 2001, mastermind suspect Khalid Sheikh Mohammed and Umar Farouk Abdulmutallab, the man accused of attempting to blow up an airplane on Christmas Day 2009.

In: Federal Protective Service
Editor: Emilius M. Zamaco
ISBN: 978-1-62417-205-2

Chapter 2

FEDERAL PROTECTIVE SERVICE: BETTER DATA ON FACILITY JURISDICTIONS NEEDED TO ENHANCE COLLABORATION WITH STATE AND LOCAL LAW ENFORCEMENT*

United States Government Accountability Office

WHY GAO DID THIS STUDY

The Department of Homeland Security's (DHS) Federal Protective Service (FPS) protects over 9,000 federal facilities under the custody and control of the General Services Administration (GSA). In 2007, FPS adopted an inspector-based workforce approach and indicated it would increase its reliance on state and local law enforcement agencies to respond to incidents at these facilities. These facilities range from facilities of proprietary or concurrent jurisdiction— where authority is shared by federal and state and local police—to facilities of exclusive jurisdiction, where only federal law enforcement has authority. As requested, this chapter assesses FPS's efforts to collaborate with state and local law enforcement for assistance in responding to incidents at these federal facilities. GAO reviewed documents on

---

* This is an edited, reformatted and augmented version of United States Government Accountability Office Publication No. GAO-12-434, dated March 2012.

collaboration, GSA and FPS facility data, and GAO's work on key collaboration practices and internal control standards. GAO also contacted 73 selected state and local law enforcement agencies from geographic jurisdictions of varying population sizes and FPS buildings throughout the United States and interviewed FPS and GSA officials.

What GAO Recommends

In conjunction with the revised MOU that is being developed between GSA and FPS, GAO recommends the administrator of GSA ensure that efforts to identify the jurisdictions of all GSA buildings are completed and that these data are provided to FPS. GSA concurred with the recommendation.

What GAO Found

To collaborate with state and local law enforcement, the Federal Protective Service (FPS) uses memorandums of understanding (MOU), long-standing working relationships, written guidance to FPS staff, joint operations, and other initiatives. For example, FPS has MOUs ranging from sharing radio frequency usage in Alabama, to a mutual aid agreement with the Metropolitan Atlanta Rapid Transit Authority in Georgia. In some jurisdictions, such as the suburbs of the District of Columbia, FPS has no MOUs but has regular contact and longstanding mutual aid relationships with state and local law enforcement. To collaborate with state and local law enforcement, FPS has guidance that addresses issues such as the scope of law enforcement authorities on federal property and information sharing among jurisdictions. FPS established regional staff positions intended to improve collaboration with other organizations and has engaged in joint operations with state and local law enforcement. By comparison, other federal organizations with law enforcement responsibilities similar to FPS also use a variety of methods, ranging from the Department of Veterans Affairs' policy to seek MOUs with state and local law enforcement to the Smithsonian Institution's established relationships with the Metropolitan D.C. Police and others.

GAO found that state and local law enforcement organizations it contacted are generally willing to assist with incidents at federal facilities. For example, 48 of 52 respondents from state and local law enforcement agencies GAO contacted about this issue said that they would respond to a call at a federally owned facility; 27 said they had done so since 2007. Overall, the variety of efforts FPS has under way is consistent with the key collaboration practices GAO has previously identified and reflects a reasonable approach to collaboration, especially when combined with the willingness of state and local law enforcement to assist.

Although FPS has a reasonable approach to state and local collaboration, GAO found issues related to the quality of data exchanged between GSA and FPS on buildings and their locations. Through working groups, GSA is working with FPS to address these data inconsistency issues and is establishing a permanent GSA liaison at FPS's headquarters to improve data coordination. But as of the end of GAO's review, FPS still lacked complete data from GSA on the jurisdiction of about one third of the buildings it protects. GSA officials informed GAO that they are making progress with identifying building jurisdictions but were not yet in a position to provide complete information to FPS. These data are important because state and local law enforcement generally has no authority to enforce state and local law on properties of exclusive federal jurisdiction. An additional effect of not having these data is that FPS lacks assurance that in relying on state and local law enforcement to respond to incidents at federal facilities, it is not creating a situation where these entities may be exercising police authority where they lack such authority. As a result, incomplete jurisdictional data leaves FPS and state and local law enforcement less equipped to define and agree to respective roles and responsibilities when there are incidents at federal facilities.

ABBREVIATIONS

ADFO	Assistant Director for Field Operations
DHS	Department of Homeland Security
FPS	Federal Protective Service
GSA	General Services Administration
MOU	memorandums of understanding
MPD	Metropolitan Police Department
NPS	National Park Service
SI	Smithsonian Institution
VA	Department of Veterans Affairs
VAMC	Department of Veterans Affairs Medical Centers

The Honorable Bennie G. Thompson
Ranking Member
Committee on Homeland Security
U.S. House of Representatives

Dear Mr. Thompson:

The Department of Homeland Security's (DHS) Federal Protective Service (FPS) protects over 9,000 federal facilities under the custody and control of the General Services Administration (GSA) by delivering integrated law enforcement and security services. To secure these buildings and safeguard their occupants, FPS employs about 1,225 federal staff, including law enforcement officers, investigators, and administrative personnel, and is responsible for management and oversight of approximately 14,000 contract security guards. These federal facilities range from facilities of proprietary or concurrent jurisdiction—in which authority is shared by federal and state and local police—to facilities of exclusive jurisdiction in which only federal law enforcement has authority. FPS conducts its mission by providing security services through two types of activities: (1) physical security activities, such as conducting risk assessments of GSA facilities and recommending risk-based countermeasures to GSA and tenant agencies aimed at reducing the likelihood and severity of incidents at facilities, and (2) law enforcement activities, such as responding to incidents, conducting criminal investigations, and exercising arrest authority. In order for FPS to track the continuous changes in the inventory of buildings that it protects—including new construction, disposed buildings, and over 8,000 leases—GSA, through its Public Buildings Service,[1] provides building data to FPS weekly from GSA's building property system.

In 2007, FPS adopted an inspector-based workforce approach to protect GSA-controlled facilities.[2] In testimony to Congress that same year, a DHS senior official stated that the agency was increasing its reliance on state and local law enforcement agencies to assist in responding to law enforcement incidents at these facilities as a way of ensuring that GSA buildings are adequately protected.[3] While FPS is statutorily responsible for enforcing federal laws and regulations for the protection of federal property and persons located on federal property regardless of the location, state and local law enforcement agencies are responsible for enforcing state laws and local ordinances within the particular state on concurrent and proprietary jurisdiction locations. In 2008, we recommended that FPS clarify roles and responsibilities of local law enforcement agencies in responding to incidents at GSA-controlled facilities.[4] Since this chapter was issued, FPS has issued specific agency instructions regarding coordination, hired new personnel to oversee coordination in the regions, and pursued additional memorandums of understanding (MOU)[5] when needed, such as in the case of clarifying roles and responsibilities between FPS and state and local law enforcement.

You asked us to provide an update on these issues. This chapter assesses FPS's efforts to collaborate with state and local law enforcement for assistance in responding to incidents at federal facilities. To meet this objective, we reviewed relevant documentation on federal facility building data from GSA. In addition, we emailed a self-administered set of 22 structured questions to the heads of 73 state and local law enforcement agencies in jurisdictions throughout the United States that we determined had FPS-protected buildings. Our non-random selection of locations included varying population sizes served by the state and local law enforcement agencies located in a mixture of urban and rural areas as defined by the United States Department of Agriculture. We used the most recent Rural-Urban Continuum Codes.[6] We followed up our emails with phone calls to these state and local law enforcement agencies to obtain clarification of the responses as needed. Not every respondent answered every question. Although the results of our structured questionnaires cannot be generalized to the universe of jurisdictions that interact with FPS, the results provide key insights and illustrate how these organizations can help FPS respond to incidents.

Further, we interviewed GSA officials at GSA Headquarters in Washington, D.C. We also interviewed FPS officials at the Suitland, Maryland, and Battle Creek, Michigan, FPS MegaCenters 7 and toured the Suitland MegaCenter. We interviewed each of the 11 FPS Regional Directors to determine how their respective regions coordinate with state and local law enforcement agencies for the properties in their jurisdiction. For comparison with FPS, we contacted three federal agencies that provide their own law enforcement services at their facilities—the Department of Veterans Affairs, the Department of the Interior's National Park Service, and the Smithsonian Institution. We obtained relevant documents pertaining to their collaboration with state and local law enforcement and interviewed key officials. Lastly, we reviewed prior GAO work, including reports on key practices in interagency collaboration, facility protection, and internal controls. We also reviewed FPS data on buildings protected, staffing, procedures, and MOUs that are in place and assessed the quality and completeness of these data. Collectively, this multifaceted approach enabled us to make conclusions about whether FPS's collaboration approach is reasonable. Appendix I contains more information on our objective, scope, and methodology.

We conducted this performance audit from February 2011 to March 2012 in accordance with generally accepted government auditing standards. Those standards require that we plan and perform the audit to obtain sufficient, appropriate evidence to provide a reasonable basis for our findings and conclusions based on our audit objectives. We believe that the evidence obtained provides a reasonable basis for our findings and conclusions based on our audit objectives.

BACKGROUND

FPS—located within the National Protection and Programs Directorate[8] of DHS—protects the over 9,000 federal facilities that are under the control and custody of GSA, as well as the persons on those properties. FPS headquarters is located in Washington, D.C.; regional offices are located in New York, Boston, Philadelphia, Atlanta, Denver, Chicago, San Francisco, Seattle, Fort Worth, Kansas City, and the District of Columbia. FPS is authorized to enforce federal laws and regulations aimed at protecting GSA buildings and persons on the property and to investigate offenses against these buildings and persons.[9] These federal facilities include buildings of exclusive, concurrent, and proprietary jurisdictions.

- Exclusive: Under exclusive federal jurisdiction, the federal government—and federal law enforcement entities—have all of the legislative authority within the land area in question, while the state—and its state and local law enforcement entities—have no residual police powers.
- Concurrent: In concurrent jurisdiction facilities, both federal and state governments—and law enforcement entities—have jurisdiction over the property.
- Proprietary: Under proprietary jurisdiction, the federal government has rights—similar to a private landowner—but also maintains its authorities and responsibilities as the federal government. Under proprietary jurisdiction, the local government is the principal municipal police authority.

To enable FPS to track changes in the inventory of federal buildings that it protects, GSA, through its Public Buildings Service, provides building data in electronic files to FPS weekly from GSA's building property system. These data include each building number with address; type of jurisdiction; and square footage and number of personnel to assist FPS to bill for its services, among other things. FPS personnel then input this information into its systems electronically. MegaCenters—the four regional dispatch centers within FPS that are the primary focal points for incident notification (see figure 1)—use the data to direct calls concerning building incidents and emergencies to FPS personnel as well as state and local law enforcement agencies.

Source: GAO.

Figure 1. Federal Protective Service MegaCenter, Suitland, Maryland.

As a general practice, MegaCenters make direct radio calls for incident response to FPS personnel and telephone calls to state and local law enforcement agencies. FPS instructs tenants to contact the MegaCenter by calling 1-877-4FPS-411 and, in areas where FPS responders cannot provide an immediate response; tenants are often directed to also dial 911. (See fig. 2.)

In this chapter, we assess GSA and FPS's processes against GAO's collaboration key practices and internal control standards. Our previous work has broadly defined collaboration as any joint activity that is intended to produce more public value than could be produced when organizations act alone.[10] We have found that key practices for collaboration include:

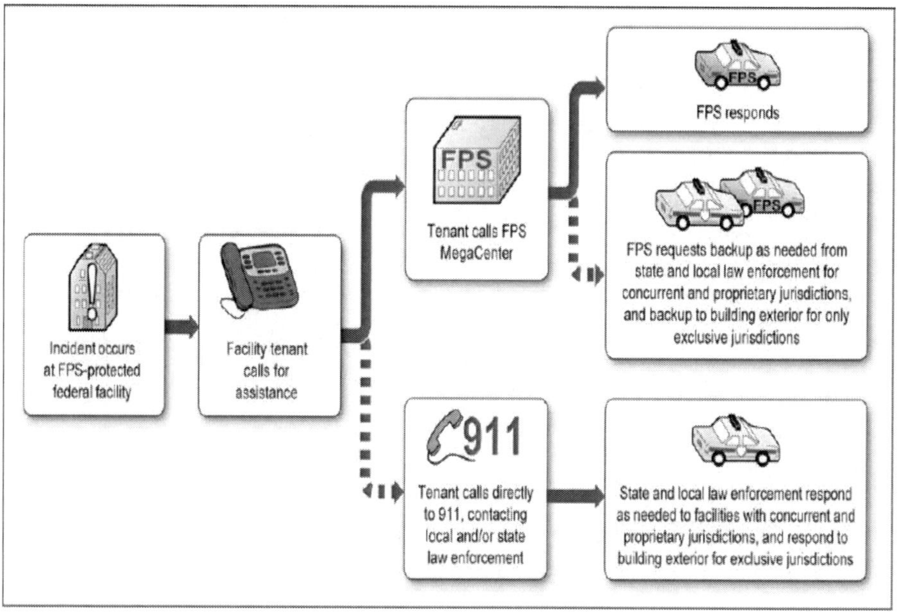

Source: GAO.

Figure 2. Notification and Response Actions Following an Incident at an FPS-Protected Facility.

- identifying and addressing needs by leveraging resources to support the common outcome and, where necessary, identifying opportunities to leverage resources;
- agreeing upon agency roles and responsibilities; and
- establishing compatible policies, procedures, and other means to operate across agency boundaries.

Additionally, agencies can strengthen their commitment to work collaboratively by articulating their agreements in formal documents, such as an MOU. We have also reported that organizations may face a range of barriers when they attempt to collaborate with other organizations. One such barrier stems from agencies' concerns about protecting jurisdiction over missions. In addition, interagency collaboration is often hindered by incompatible procedures, processes, data, and computer systems. GAO has identified standards in facility protection that provide a framework for guiding agencies' efforts in this area, such as establishing a means of coordinating and sharing information with other government entities and the private sector.[11] Finally, standards for controls over information processing come from GAO's *Standards for Internal Control in the Federal Government*. According to these standards, internal control is a major part of managing an organization and comprises the plans, methods, and procedures used to meet missions, goals, and objectives.[12] Internal control standards specific for information systems help ensure the completeness and accuracy of data.

Some other federal agencies provide their own law enforcement at their facilities. These include the Department of Veterans Affairs (VA) Police, National Park Service (NPS) law enforcement within the Department of the Interior, and Smithsonian Institution (SI) Police. VA Police provide law enforcement duties to the 152 VA Medical Centers (VAMC). [13] U.S. Park Police is a unit of the National Park Service, with jurisdiction in all National Park Service areas and certain other federal and state lands.[14] U.S. Park Police provides law enforcement services to designated areas within the National Park Service (primarily in the District of Columbia, New York City, and San Francisco, California metropolitan areas). Additionally, Law Enforcement Park Rangers, belonging to the "Visitor and Resource Protection Division" within the National Park Service, are authorized to carry firearms, conduct investigations, make arrests, and serve warrants pursuant to law and policy. Protection and security services at Smithsonian Institution facilities are provided by the Smithsonian Police.[15]

FPS Uses Various Approaches to Collaborate with State and Local Law Enforcement, but Jurisdictional Data on Federal Facilities Are Incomplete

FPS's Approach Involves a Variety of MOUs, Reliance on Long-standing Relationships, Guidance, and Other Initiatives

FPS uses a variety of methods to collaborate with state and local law enforcement, ranging from establishing MOUs to document agreement on roles and responsibilities with some, to relying on long-standing working relationships with others. FPS also has guidance and various other efforts under way related to coordination with state and local law enforcement. More specifically, FPS reported that it had 21 signed MOUs with state and local law enforcement agencies across the United States as of September 2011.[16] For example, there is an MOU for radio frequency usage in Alabama; MOUs for arrest authority on properties adjacent to federal property in California and Florida; and MOUs for mutual aid in the District of Columbia and Georgia, such as FPS's reciprocal support agreement with Metropolitan Atlanta Rapid Transit Authority. MOUs are mechanisms that can be used to formalize key practices in agency collaboration such as agreeing on roles and responsibilities, including leadership, and to establish compatible policies, procedures, and other means to operate across agency boundaries.[17]

While FPS has often found MOUs helpful, the general consensus among FPS officials was that effective coordination did not depend on having MOUs. FPS prefers a flexible approach of pursuing MOUs only when it determines they are needed, as opposed to seeking them in all cases. FPS's Director stated that FPS has generally not found it necessary to create written documents, requirements, or MOUs because FPS has always received good cooperation from state and local law enforcement agencies when their assistance was needed. For example, in some jurisdictions such as the suburbs surrounding the District of Columbia, FPS has no MOUs with state and local law enforcement agencies but has regular contact and longstanding mutual aid relationships. In addition, several FPS Regional Directors highlighted the importance of local response to incidents in and around federal facilities in rural areas because of the lack of FPS staff at these locations and noted that their informal relationships have worked successfully because state and local law enforcement agencies were consistently reliable in their response to these locations. FPS officials stated that mandating the pursuance of MOUs with all law enforcement entities would not be in the best interest of effectiveness and efficiency and would increase the burden on already task-saturated FPS staff. In addition, it is generally up to state and local law enforcement agencies as to whether they would be willing to enter into an MOU with a federal agency.

With regard to long-standing working relationships and regular contact with state and local law enforcement, FPS Inspectors and Regional Directors have developed relationships with state and local law enforcement agencies and collaborate on different levels. Regional Directors in all 11 FPS regions stated that their offices routinely had direct contact with state and local law enforcement agencies at multiple types of security meetings such as the Federal Executive Boards, joint terrorism task forces, and regional fusion center meetings.[18] Attendance by both state and local law enforcement agencies and FPS at these meetings establishes mutually reinforcing or joint strategies designed to help align activities, core processes, and resources to achieve a common outcome. For example, FPS participates in monthly meetings of the Law Enforcement Working Group of the Atlanta Downtown Improvement District. State and local law enforcement chiefs or deputy chiefs from the surrounding area, officials from 15 local colleges, and officials from other federal agencies participate. According to an FPS regional official, the group acts as a "force multiplier" to fight crime within the district, which includes GSA-controlled facilities. FPS officials also have discussions with state and local law enforcement agencies as needed during operational planning associated with special events such as the Olympics, protests, and parades.

FPS also has guidance for FPS staff and other efforts under way to collaborate with state and local law enforcement. Regional Directors are responsible for carrying out FPS policy and guidance, and state that many of these written policies contain directives for collaboration with state and local law enforcement. One such directive is FPS Directive 15.1.2.1, Law Enforcement Authority and Powers, which outlines the scope of law enforcement authority on federal property. Other policies that reference state and local law enforcement agencies' coordination include FPS's Regional Information Sharing Program, Detention and Arrest, and Joint Terrorism Task Force Policy, among others. Best practices and lessons learned are also communicated throughout FPS regions with weekly regional director conference calls, a regional director's council that meets monthly, and yearly Regional Director conferences.

In addition to guidance, FPS added three new positions in fiscal year 2011 intended to improve communication and standardization across FPS by coordinating with federal, state, and local law enforcement officials to reduce crime at, and potential threats to, federal facilities. These positions are titled Assistant Director for Field Operations (ADFO) for west, central, and east operations. The ADFO will be a spokesman for FPS, representing the Director in his or her designated area.

Other initiatives employed by FPS include collaborative operations to avert or obstruct potential threats inside the facility, such as the presence of unauthorized persons, or potentially disruptive or dangerous activities, such as potential terrorist operations and criminal activity in and around federal buildings. Using a combination of law enforcement agencies is consistent with facility protection key practices to establish a means of coordinating and sharing information with other government entities. These operations begin with planning meetings involving FPS and any other federal, state, and local law enforcement agencies that may be called upon to assist. The operations combine physical security expertise and law enforcement authority into an enhanced security team to provide a visual deterrent at FPS-protected facilities. The combined team then selects a federal building for which FPS has security oversight and provides a highly visible presence for a select period of time with patrol operations, explosive detection dog sweeps, and an enhanced security posture. As a means to leverage resources, FPS has collaborated with state and local law enforcement to assist in conducting these operations by enlisting their support in Chicago, Illinois; New York, New York; Newark, New Jersey; and the District of Columbia.

Like FPS, Other Federal Agencies Use a Variety of Methods to Collaborate with State and Local Law Enforcement

Other federal organizations with law enforcement responsibilities similar to FPS also used a variety of methods to collaborate with state and local law enforcement. For example, VA has a policy requiring all locations of VA-controlled property to have formally documented MOUs with state and local law enforcement agencies to ensure timely backup support for VA Police officers.[19] VA headquarters officials stated that MOUs are useful because VA Police typically transport detainees to state and local law enforcement agencies for arrest and processing, while state and local law enforcement agencies typically provide first response to leased property under VA control. However, VA Police reported they cannot provide mutual aid to state and local law enforcement agencies on non-VA controlled property because existing law[20] limits the authority of Department police officers to VA property.

According to U.S. Park Police headquarters officials, the Park Police has MOUs with federal, state, and local law enforcement including a longstanding formal relationship with the District of Columbia Metropolitan Police Department (MPD). Some of the MOUs are for events and are short-term, such as the last presidential inauguration in the District of Columbia. The U.S. Park Police also stated they have MOUs that are formal incident response plans, which outline the roles and responsibilities of the various entities. In the District of Columbia, the U.S. Park Police responded to the U.S. Holocaust Museum shooting incident in 2009 and have provided service to the Kennedy Center for the Performing Arts for a fee.

According to a Smithsonian Institution headquarters official, the Smithsonian police rely on MPD and the Park Police to arrest and process suspects in the District of Columbia. Aside from a limited MOU with the

District of Columbia, MPD[21] the Smithsonian police does not have MOUs with other law enforcement agencies. In the District of Columbia region as well as nationally, Smithsonian police rely on informal relationships with state and local law enforcement agencies for coordination of law enforcement at Smithsonian-controlled property. However, Smithsonian police officials stated they cannot provide mutual aid— due to lack of authority—to state and local law enforcement agencies on non-Smithsonian-controlled property.

State and Local Law Enforcement Generally Willing to Assist with Protecting Federal Facilities

Our questionnaire of state and local law enforcement agencies and follow-up discussions showed a general willingness of those that replied to respond to incidents at federal facilities. For example, 48 of 52 agencies that answered the question replied that they would respond to calls that dispatch to a federally occupied (owned and/or leased) building, and 27 of 44 had actually responded to a federally occupied building since 2007. As for MOUs, 11 of 43 agencies that answered the question reported having formal MOUs with FPS and 4 of 40 reported having informal agreements. (See table 1.) Four state and local law enforcement agencies stated that they would decline to respond to an incident at a federal building in their jurisdiction. Three of these four law enforcement agencies were sheriff or highway patrol entities that stated that they are not the first responders to incidents at the facilities in question and that there were local police available for response. A fourth questionnaire responder did not clarify why it answered negatively; however, additional inquiry with the federal property owner in this law enforcement's jurisdiction stated the particular law enforcement agency did coordinate and respond to calls at the property. Only one state and local law enforcement agency replied that it was denied access to a federal building when responding to an incident within its jurisdiction; however, it declined to clarify the specific instance in which it was denied access. The only law enforcement agency that answered it had declined to respond to a call dispatched at a federally occupied (owned and/or leased) building, later clarified that the answer applied to non-GSA-controlled facilities such as buildings of Department of Defense and other federal agencies. Table 1 shows the specific questions and responses provided by state and local law enforcement.

Table 1. State and Local Law Enforcement Agencies' Responses to Selected Questions in GAO's Structured Questionnaire Regarding FPS Collaboration

	Answers		
Questions[a]	**Yes**	**No**	**No answer**
To your knowledge, are there federal buildings located within your jurisdiction?[b]	55	1	1
Would your law enforcement entity respond to calls that dispatches to a federally occupied (owned and/or leased) building?	48	4	5
Since 2007, has your law enforcement entity been denied access into federally occupied (owned and/or leased) building in response to an incident under your entity's purview (proprietary of the state or local government, or concurrent with the federal government)?	1	38	18
Since 2007, has your law enforcement entity responded to a call that dispatches to a federally occupied (owned and/or leased) building?	27	17	13
Since 2007, has your law enforcement entity declined to respond to a call that dispatches to a federally occupied (owned and/or leased) building?	1	43	13
Since 2007, does your law enforcement entity have any formal memorandums of understanding (MOUs) with the U.S. Department of Homeland Security, Federal Protective Service?	11	32	14
Does your law enforcement entity have any informal agreements with the U.S. Department of Homeland Security, Federal Protective Service?	4	36	17

Source: GAO structured questionnaire of state and local law enforcement agencies.
[a]llot every respondent answered every question.
[b]Although we selected our respondents based on addresses of federal buildings located in their probable geographic jurisdictions, upon further examination, one respondent noted that a building address we chose was actually one street out of its jurisdiction. Therefore it answered that it had no federal buildings in its area of jurisdiction.

FPS's Approach Is Reasonable and Consistent with Key Practices

Overall, the variety of efforts FPS has under way reflects a reasonable approach to collaboration, especially when combined with results we found from our questionnaire of state and local law enforcement agencies. The practice of maintaining working relationships and having regular contact with state and local law enforcement officials establishes mutually reinforcing or joint strategies designed to help align activities, core processes, and resources to achieve a common outcome. The MOUs that FPS has in place are mechanisms consistent with facilitating key practices in agency collaboration, such as defining and agreeing to roles and responsibilities. Establishing compatible policies, procedures, and other means to operate across agency boundaries are key practices that can help enhance and sustain collaboration. Pursuing MOUs on an as-needed basis is also consistent with how other federal law enforcement agencies approach collaboration. Performing operations such as extra patrol activities using a combination of law enforcement agencies is consistent with facility protection key practices to establish a means of coordinating and sharing information with other government entities.

Missing Jurisdictional Data Are of Concern, and Data Inconsistencies with GSA Are Being Addressed

Although FPS's approach to collaboration is reasonable, issues related to data quality arose during our review. Specifically, we found that FPS lacked complete data from GSA on the type of jurisdiction (e.g., concurrent or exclusive) for about one-third of the buildings FPS protects, making it difficult to ensure that it is addressing the full scope of issues related to jurisdictional roles and responsibilities. At the end of our review, GSA officials informed us that they had made significant progress addressing this issue. More specifically, when we reviewed the property list that GSA provided to FPS in December 2011—which is provided on a weekly basis—about thirty-four percent of the properties lacked recorded jurisdictions, including blank and pending jurisdiction categories. (See table 2.)

GSA officials stated that they were aware of the numerous blank data fields pertaining to jurisdictions and that they were trying to individually assess these fields building by building. They further stated that it was a time-consuming process that included reviewing individual property historical records. GSA officials stated they had made progress and the jurisdictions that have not been identified were down to 2 percent. However, these data had not yet been added to GSA's building property system or contained in the electronic files GSA sends to FPS weekly. GSA officials also stated the jurisdictional field on the GSA property list was not in the top fifty fields that the agency typically monitors because of the large number of data fields, although the officials recognized the importance of this field to FPS.

Table 2. Jurisdiction of Properties under the Control of GSA with FPS Responsibility

Jurisdiction[a,b]	Number of GSA Properties	Percentage
Pending	1594	18
Exclusive	329	4
Concurrent	662	7
Partial	31	0
Proprietary	4969	55
Combined	21	0
Blank[b]	1432	16
Total	9038	100

Source: GAO analysis of GSA data.

[a]GSA provides the following guidance when assigning jurisdiction to its properties: Pending: requires value assignment in the database; Exclusive: the U.S. enjoys exclusive legislative jurisdiction and the state has no legislative jurisdiction except for minor forms of taxation; Concurrent: both the U.S. and a state enjoy complete legislative jurisdiction; Partial: the state has selectively ceded certain aspects of legislative jurisdiction to the U.S; Proprietary: the state exercises complete legislative jurisdiction and the U.S. is considered only a property holder; Combined: any location at which more than one of the above jurisdictions exist.

[b]Blank represents number of facilities missing GSA jurisdictional data.

During our review, we found no instances in which state or local law enforcement exceeded their jurisdictional authority. In some instances, state and local law enforcement responded to the perimeter of buildings with exclusive jurisdiction for matters such as traffic accidents and suspicious packages. FPS officials said that state and local law enforcement may also be granted access if officers are in pursuit of a suspect. Furthermore, FPS officials said that inspectors and GSA staff at the building level generally know the jurisdiction of the individual buildings for which they are responsible.

Nonetheless, given that facilities of exclusive jurisdiction are unique because state and local law enforcement agencies generally have no law enforcement authority on these properties, incomplete data leaves FPS less equipped to define and agree to respective roles and responsibilities with regard to state and local law enforcement collaboration. An additional effect of not having these data is that FPS lacks assurance that, in relying on state and local law enforcement to respond to incidents at federal facilities, it is not creating a situation where these entities may be exercising police authority where they lack such authority as in the case of exclusive jurisdiction properties. In addition, having incomplete data is inconsistent with established standards for internal control over data systems, including those standards that relate to accuracy and completeness. While only 4 percent of GSA's inventory was known to be of exclusive jurisdiction, 34 percent of GSA's inventory had incomplete data on the type of jurisdiction in GSA's building property system.

In our review, we also found inconsistencies between FPS and GSA data on buildings and their locations—6 of the 11 FPS regions reported that the GSA list does not match the current property inventory. One FPS regional official stated that GSA does not keep the property list as current as FPS needs; changes occur but are not captured by GSA. For example, the official stated that in his region, agencies sign leases about a dozen times a year without FPS's knowledge or timely notice. FPS officials noted that the overall number could be greater across all FPS regions. The current MOU between GSA and FPS calls for a pre-lease assessment of the building by FPS, but these assessments cannot be completed if FPS is unaware of the new lease.[22] Another FPS regional official stated that the region uses its own building list, which is updated by FPS regularly as information becomes available. A third FPS official stated that the GSA list does not capture changes to buildings with a security risk level of 1 or 2 as quickly as FPS needs.[23] A fourth regional official stated that the region relies on a combination of building lists from GSA, FPS provided lists, and its own regional list. This official stated these lists often do not reconcile because of changes that are not updated in a timely manner. In addition, a majority of state and local law enforcement agencies we sent questions to replied that they did not identify the jurisdiction of the individual federal buildings in their geographic areas, while three entities replied that they only identified some building jurisdictions.

GSA officials recognize that the exchange of building data with FPS is an issue. GSA stated that only recently did it have the ability to cross-reference and address these differences, and is working with FPS to correct them. For example, in 2011, GSA and FPS held working groups to begin to improve the building property list, and established a permanent GSA liaison in FPS's headquarters to improve data coordination. Although this effort is still in progress and data inconsistencies remain, GSA and FPS are addressing concerns about data inconsistencies. Further, GSA and FPS are currently negotiating a new MOU that is expected to be finalized in early 2012. GSA officials told us that the new MOU will include an agreement on sharing information, such as the building data, and specifically sharing information at the regional level. FPS and GSA did not indicate whether the revised MOU would address the aforementioned issue related to incomplete jurisdictional data. However, it would seem that addressing this issue in conjunction with revising the MOU would ensure that data shared were not only consistent, but more complete as well.

CONCLUSION

FPS's approach to collaborating with state and local law enforcement is reasonable and consistent with key practices in that the approach uses mechanisms such as MOUs to document agreements on roles and responsibilities in some cases, long-standing working relationships, written guidance to FPS staff, joint operations, and other initiatives to promote mutual collaboration. Other federal organizations with law enforcement responsibilities similar to FPS—such as VA, U.S. Park Police, and Smithsonian—also use a variety of methods for collaboration with state and local law enforcement. State and local law enforcement agencies we contacted were generally willing to assist FPS with incidents at federal facilities. Related to the quality of data exchanged between FPS and GSA on buildings and their locations, FPS and GSA had taken action to address data inconsistency issues. However, as of the end of our review, FPS still lacked complete data from GSA on whether the jurisdictions of about one-third of the buildings FPS protects are exclusive, concurrent, or proprietary. Having these data is important because state and local law enforcement generally have no authority to enforce state and local law on properties of exclusive jurisdiction. At the end of our review, GSA informed us that it had made progress with addressing this issue. GSA and FPS are negotiating a revised MOU that will include agreement on sharing information such as the building data. As such, addressing the issue related to incomplete data on jurisdictions, in conjunction with revising the MOU, would ensure that data were not only consistent, but more complete as well. Otherwise, FPS would remain less equipped to define and agree to respective roles and responsibilities as it proceeds with its efforts to rely on state and local law enforcement for assistance in responding to incidents at federal facilities.

RECOMMENDATIONS FOR EXECUTIVE ACTION

In conjunction with the revised MOU that is being developed between GSA and FPS, we recommend the Administrator of GSA ensure that efforts to identify the jurisdictions of all GSA buildings are completed and that these data are provided to FPS so that FPS is better equipped to manage jurisdictional roles and responsibilities at GSA buildings.

AGENCY COMMENTS AND OUR EVALUATION

We provided a draft of this chapter to GSA, DHS, DOI, VA, and Smithsonian Institution for their review and comment. GSA concurred with our recommendation that the Administrator of GSA ensure that efforts to identify the jurisdictions of all GSA buildings are completed, and that these data are provided to FPS so that FPS is better equipped to manage jurisdictional roles and responsibilities at GSA buildings. DHS provided a letter, describing its efforts to collaborate with state and local law enforcement. DHS also provided technical comments, which we incorporated, as appropriate. DOI, VA, and the Smithsonian Institution provided minor technical comments, via email, which we incorporated, as appropriate.

Sincerely yours,
Mark Goldstein Director
Physical Infrastructure Issues

APPENDIX I: OBJECTIVE, SCOPE AND METHODOLOGY

To assess the Federal Protective Service's (FPS) efforts to collaborate with state and local law enforcement for assistance in responding to incidents at federal facilities, we reviewed FPS data on buildings protected, staffing, procedures, and memorandums of understanding (MOUs). We also reviewed relevant federal facility building data from the General Services Administration (GSA) including for example, each building number with address; type of jurisdiction; and square footage and number of personnel, among other things. We interviewed FPS officials throughout the regions, and FPS and GSA officials at their respective agency's headquarters in Washington, D.C., regarding the processes and procedures for exchanging these data. We reviewed the building data for completeness, but did not verify the accuracy of the information contained for each building. To ensure we were assessing the exact data that FPS uses, we requested data samples for fiscal year 2011 from both GSA and FPS and replicated the jurisdiction category results. We assessed the extent to which there were missing jurisdiction assignments by reviewing pending and blank jurisdiction categories. We then assessed GSA and FPS's processes for managing these data against GAO's *Standards for Internal Control in the Federal Government, Homeland Security: Further Actions Needed to Coordinate Federal Agencies' Facility Protection Efforts and Promote Key Practices, and Results-Oriented Government: Practices That Can Help Enhance and Sustain Collaboration among Federal Agencies.* [1] According to GAO's standards for internal control in the federal government, internal control is a major part of managing an organization and comprises the plans, methods, and procedures used to meet missions, goals, and objectives. Internal control, which is synonymous with management control, helps government program managers achieve desired results through effective stewardship of public resources. Control activities—such as reconciliations performed to verify data completeness; an agency's data entry design features contribute to data accuracy; data validation and editing performed to identify erroneous data; and erroneous data that is captured, reported, investigated, and promptly corrected—contribute to data accuracy and completeness. We determined that the data were sufficiently reliable for the purposes of this chapter.

For comparison with FPS's coordination efforts, we contacted three federal agencies that provide law enforcement at their facilities—the Department of Veterans Affairs (VA), the National Park Service (NPS) within the Department of the Interior, and the Smithsonian Institution (SI). To gain insight into FPS, VA, SI, and NPS coordination with state and local law enforcement agencies, we emailed a self-administered set of 22 structured questions to the heads of 73 state and local law enforcement agencies. Our non-random selection of locations included varying population sizes located in a mixture of metro, urban, and rural areas as defined by the United States Department of Agriculture using the most recent Rural-Urban Continuum Codes for jurisdictions that we determined had FPS, and/or VA, NPS, and SI buildings throughout the United States. The state and local law enforcement agencies we chose included a mix of police, sheriff, highway patrol agencies in each of the 11 FPS regions. We also followed up our email with phone calls to these state and local law enforcement agencies. Not every respondent answered every question related to coordination with FPS, VA Police, U.S. Park Police, and SI police. Additionally, the responses had varying levels of staff within the state and local law enforcement organization reply for the organization. Furthermore, the structured questions were related to coordination with the Federal Protective Service, Veterans Affairs Police, Smithsonian police, and the U.S. Park Police. Although the results of our questions cannot be generalized to the universe of jurisdictions that have interaction with FPS, the results provide key insights on how state and local law enforcement collaborates with FPS to assist with federal facility protection. These results illustrate how FPS relies on these organizations to respond to incidents and collectively, how this multi-faceted approach enabled us to make conclusions whether FPS's approach is reasonable.[2]

Further, we interviewed officials at two FPS MegaCenters—the four regional dispatch centers within FPS that are the primary focal points for initial incident notification—and toured the Suitland, Md., MegaCenter facility. We attended an FPS operational exercise in the District of Columbia. We also interviewed each of the 11 FPS Regional Directors to determine how their region coordinates with state and local law enforcement entities for the properties in their jurisdiction. We interviewed GSA officials at GSA headquarters in the District of Columbia. We obtained relevant documents pertaining to VA, NPS, and SI collaboration with state and local law enforcement and interviewed agency officials. Lastly, we reviewed prior GAO work, including reports on key practices in interagency collaboration and facility protection.

We conducted this performance audit from February 2011 to March 2012 in accordance with generally accepted government auditing standards. Those standards require that we plan and perform the audit to obtain sufficient, appropriate evidence to provide a reasonable basis for our findings and conclusions based on our audit objectives. We believe that the evidence obtained provides a reasonable basis for our findings and conclusions based on our audit objectives.

End Notes

[1]The Public Buildings Service acquires space on behalf of the federal government through new construction and leasing and acts as a caretaker for federal properties across the country.

[2]This model was intended to make more efficient use of FPS's declining staffing levels by increasing focus on FPS's physical security duties and consolidating law enforcement activities. FPS's goal was to shift its law enforcement workforce composition from a mix of about 40 percent police officers, about 50 percent inspectors, and about 10 percent special agents, to a workforce primarily composed of inspectors and some special agents. FPS's inspectors are responsible for law enforcement and security duties, including: patrolling building perimeters, responding to incidents, completing risk assessments for buildings, recommending security countermeasures, and overseeing the contract security workforce.

[3]Hearing on "Proposals to Downsize the Federal Protective Service and Effects on the Protection of Federal Buildings," Committee on Transportation and Infrastructure, U. S. House of Representatives, April 18, 2007.

[4]GAO, *Homeland Security: The Federal Protective Service Faces Several Challenges That Hamper Its Ability to Protect Federal Facilities*, GAO-08-683 (Washington, D.C.: June 11, 2008).

[5]For consistency, we are using the term MOU throughout this report for both memorandums of understanding (MOUs) and memorandums of agreement (MOAs).

[6]Rural-Urban Continuum Codes form a classification scheme that distinguishes metropolitan (metro) counties by the population size of their metro area, and nonmetropolitan (nonmetro) counties by degree of urbanization and adjacency to a metro area or areas. See USDA Economic Research Service at http://www.ers.usda.gov/Briefing/Rurality/RuralUrbCon/.

[7]In 2000, the Federal Protective Service (FPS) transitioned all alarm-monitoring and dispatching capabilities from several regional control centers to four MegaCenters. Currently, each MegaCenter monitors multiple types of alarm systems, closed circuit television, and wireless dispatch communications within federal facilities throughout the nation. These centers—located in Michigan, Colorado, Pennsylvania, and Maryland—are equipped with state-of-the-art communication systems and operate continuously.

[8]GAO, *Federal Protective Service: Progress Made but Improved Schedule and Cost Estimate Needed to Complete Transition,* GAO-11-554 (Washington, D C.: July 15, 2011).

[9]Section 1315(a) of Title 40, United States Code, provides that: "To the extent provided for by transfers made pursuant to the Homeland Security Act of 2002, the Secretary of Homeland Security...shall protect the buildings, grounds, and property that are owned, occupied, or secured by the Federal Government (including any agency, instrumentality, or wholly owned or mixed-ownership corporation thereof) and the persons on the property."

[10]GAO, *Results-Oriented Government: Practices That Can Help Enhance and Sustain Collaboration among Federal Agencies*, GAO-06-15 (Washington, D.C.: Oct. 21, 2005).

[11]GAO, *Homeland Security: Further Actions Needed to Coordinate Federal Agencies' Facility Protection Efforts and Promote Key Practices*, GAO-05-49 (Washington, D.C.: Nov. 30, 2004).

[12]GAO, *Standards for Internal Control in the Federal Government*, GAO/AIMD-00-21.3.1. (Washington, D.C.: November 1999).

[13]38 U. S. C. § 902. [14]16 U. S. C. § 1A-6(b).

[15]40 U. S. C. § 6306.

[16]We did not determine the total universe of state and local law enforcement jurisdictions where FPS-protected properties are located.

[17]GAO-06-15.

[18]The Federal Executive Boards, established by Presidential Directive in 1961, are a forum for communication and collaboration among federal agencies outside of Washington, D.C. Joint terrorism task forces (JTTFs) are small cells of highly trained, locally based investigators, analysts, linguists, special weapons and tactics experts, and other specialists from dozens of U.S. law enforcement and intelligence agencies. JTTFs are used as a multi-agency effort by the U.S. Department of Justice and the Federal Bureau of Investigation to combine the resources of federal, state, and local law enforcement. Fusion centers serve as focal points within the state and local environment for the receipt, analysis, gathering, and sharing of threat-related information between the federal government and state, local, tribal, territorial, and private sector partners.

[19]Department of Veterans Affairs: *Security and Law Enforcement VA Handbook 0730* (Washington, D.C.: Aug. 11, 2000).

[20]38 U. S. C. § 902.

[21]The MOU assists the Metropolitan Police Department. D.C. to conduct crime prevention and law enforcement activities for specific instances or periods of time specified by the Smithsonian Institution's Office of Protective Services. The MOU allows officers from the National Zoo in Washington, D.C. to direct traffic in the immediate surroundings of the Zoo or in the instances of hot pursuit for example, to take action from outside the perimeter of the Zoo if appropriate to protect the Zoo's property, visitors, and animals.

[22]The current MOU between FPS and GSA began on June 1, 2006, and was valid for 2 years. However, the MOU states that upon expiration of the initial term, the MOU will automatically be renewed and will remain in full force and effect until modified in writing, executed by both parties, or terminated by either party upon 90 days' written notice to the other party.

[23]Each federal building is assessed a security risk level, with Level 1 as lowest and Level 5 as highest. FPS uses Interagency Security Committee guidelines to determine a facility's security level, which in turn determines the level of physical protection services needed at each of the approximately 9,000 buildings.

End Notes for Appendix I

[1]GAO, *Standards for Internal Control in the Federal Government*, GAO/AIMD-00-21.3.1. (Washington, D.C.: November 1999); GAO, *Homeland Security: Further Actions Needed to Coordinate Federal Agencies' Facility Protection Efforts and Promote Key Practices*, GAO-05-49 (Washington, D.C.: Nov. 30, 2004); GAO, *Results-Oriented Government: Practices That Can Help Enhance and Sustain Collaboration among Federal Agencies*, GAO-06-15 (Washington, D.C.: Oct. 21, 2005).

[2]Results from nonprobablity samples cannot be used to make inferences about a population because in a nonprobability sample, some elements of the population being studied have no chance or an unknown chance of being selected as part of the sample.

In: Federal Protective Service
Editor: Emilius M. Zamaco

ISBN: 978-1-62417-205-2

Chapter 3

FEDERAL PROTECTIVE SERVICE: ACTIONS NEEDED TO ASSESS RISK AND BETTER MANAGE CONTRACT GUARDS AT FEDERAL FACILITIES*

United States Government Accountability Office

WHY GAO DID THIS STUDY

FPS provides security and law enforcement services to over 9,000 federal facilities under the custody and control of the General Services Administration (GSA). GAO has reported that FPS faces challenges providing security services, particularly completing FSAs and managing its contract guard program. To address these challenges, FPS spent about $35 million and 4 years developing RAMP—essentially a risk assessment and contract guard oversight tool. However, RAMP ultimately could not be used because of system problems.

GAO was asked to examine (1) the extent to which FPS is completing risk assessments; (2) the status of FPS's efforts to develop an FSA tool; and (3) FPS's efforts to manage its contract guard workforce. GAO reviewed FPS

* This is an edited, reformatted and augmented version of United States Government Accountability Office Publication No. GAO-12-739, dated August 2012.

documents, conducted site visits at 3 of FPS's 11 regions, and interviewed FPS officials and inspectors, guard companies, and 4 risk management experts.

What GAO Recommends

GAO recommends that FPS incorporate *NIPP's* risk management framework in any future risk assessment tool; coordinate with federal agencies to reduce any unnecessary duplication in FPS's assessments; address limitations with its interim tool to better assess federal facilities; develop and implement a comprehensive and reliable contract guard oversight system; and independently verify that its contract guards are current on all training and certification requirements. DHS concurred with GAO's recommendations.

What GAO Found

The Department of Homeland Security's (DHS) Federal Protective Service (FPS) is not assessing risks at federal facilities in a manner consistent with standards such as the *National Infrastructure Protection Plan's* (*NIPP*) risk management framework, as FPS originally planned. Instead of conducting risk assessments, since September 2011, FPS's inspectors have collected information, such as the location, purpose, agency contacts, and current countermeasures (e.g., perimeter security, access controls, and closed-circuit television systems). This information notwithstanding, FPS has a backlog of federal facilities that have not been assessed for several years. According to FPS's data, more than 5,000 facilities were to be assessed in fiscal years 2010 through 2012. However, GAO was unable to determine the extent of FPS's facility security assessment (FSA) backlog because the data were unreliable. Multiple agencies have expended resources to conduct risk assessments, even though the agencies also already pay FPS for this service. FPS received $236 million in basic security fees from agencies to conduct FSAs and other security services in fiscal year 2011. Beyond not having a reliable tool for conducting assessments, FPS continues to lack reliable data, which has hampered the agency's ability to manage its FSA program.

FPS has an interim vulnerability assessment tool, referred to as the Modified Infrastructure Survey Tool (MIST), which it plans to use to assess federal facilities until it develops a longer-term solution. According to FPS, once implemented, MIST will allow it to resume assessing federal facilities' vulnerabilities and recommend countermeasures—something FPS has not done consistently for several years. Furthermore, in developing MIST, FPS generally followed GAO's project management best practices, such as conducting user acceptance testing. However, MIST has some limitations. Most notably, MIST does not estimate the consequences of an undesirable event occurring at a facility. Three of the four risk assessment experts GAO spoke with generally agreed that a tool that does not estimate consequences does not allow an agency to fully assess risks. FPS officials stated that they did not include consequence information in MIST because it was not part of the original design and thus requires more time to validate. MIST also was not designed to compare risks across federal facilities. Thus, FPS has limited assurance that critical risks at federal facilities are being prioritized and mitigated.

FPS continues to face challenges in overseeing its approximately 12,500 contract guards. FPS developed the Risk Assessment and Management Program (RAMP) to help it oversee its contract guard workforce by (1) verifying that guards are trained and certified, and (2) conducting guard post inspections. However, FPS faced challenges using RAMP, such as verifying guard training and certification information, for either purpose and has recently determined that it would no longer use RAMP. Without a comprehensive system, it is more difficult for FPS to oversee its contract guard workforce. FPS is verifying guard certification and training information by conducting monthly audits of guard contractor training and certification information. However, FPS does not independently verify the contractor's information. Additionally, according to FPS officials, FPS recently decided to deploy a new interim method to record post inspections to replace RAMP.

ABBREVIATIONS

DHS	Department of Homeland Security
EPA	Environmental Protection Agency
FEMA	Federal Emergency Management Agency
FPS	Federal Protective Service
FSA	facility security assessment
FSC	Facility Security Committee
FSL	facility security level
FSRM	Federal Security Risk Manager

GSA	General Services Administration
IG	Inspector General
IP	Office of Infrastructure Protection
IRS	Internal Revenue Service
IRVS	Integrated Rapid Visual Screening of Buildings
ISC	Interagency Security Committee
IST	Infrastructure Survey Tool
MIST	Modified Infrastructure Survey Tool
NIPP	National Infrastructure Protection Plan
NPPD	National Protection and Programs Directorate
RAMP	Risk Assessment and Management Program
S&T	Science and Technology Directorate

The Honorable Daniel E. Lungren
Chairman
Subcommittee on Cybersecurity,
Infrastructure Protection,
and Security Technologies,
Committee on Homeland Security
House of Representatives

Dear Mr. Chairman:

Federal facilities are among the targets for terrorist attacks and other acts of violence, as evidenced by the 2012 shooting at the Anderson Federal Building in Long Beach, California, and the 2011 attempted bombing of the McNamara Federal Building in Detroit, Michigan. These incidents highlight the importance of protecting the over one million government employees who work in, as well as the public who visit, the more than 9,000 federal facilities under the custody and control of the General Services Administration (GSA). As a component of the Department of Homeland Security's (DHS) National Protection and Programs Directorate (NPPD), the Federal Protective Service (FPS) is the primary agency responsible for protecting these facilities. FPS provides physical security services, such as conducting risk assessments, which FPS refers to as facility security assessments (FSA), and responds to incidents at federal facilities. An FSA helps FPS identify and evaluate potential risks so that countermeasures can be recommended to help prevent or mitigate these risks.

We have previously reported that FPS faces long-standing challenges in providing security services, particularly in completing quality risk assessments in a timely manner and overseeing its contract guard program. To address challenges related to FPS's FSA process and contract guard oversight, FPS developed the Risk Assessment and Management Program (RAMP), a Web-enabled FSA and guard management system, which was implemented in November 2009. We reported in July 2011 that FPS spent about $35 million and took almost 4 years developing RAMP—$14 million and 2 years more than planned.[1] However, RAMP could not be used to complete FSAs because of several factors, including that FPS did not verify the accuracy of federal facility data used. As a result, FPS's Director decided that the agency would stop using RAMP to conduct FSAs and instead pursue an interim tool to replace it. In September 2011, FPS began working with Argonne National Laboratory to develop a vulnerability assessment tool referred to as the Modified Infrastructure Survey Tool (MIST). FPS plans to use MIST to assess the vulnerabilities of federal facilities until the agency develops a permanent replacement for RAMP. FPS also developed RAMP to (1) provide accurate and reliable records of its contract guards' training and certifications that FPS could use to verify that guards deployed at federal facilities are qualified and (2) to conduct guard post inspections. In July 2011, we reported that FPS had experienced difficulty using RAMP to ensure that its guards met training and certification requirements, primarily because of challenges with verifying RAMP's guard data.[2] FPS concurred with our recommendation to determine whether it was cost beneficial to continue to use RAMP for guard oversight. On June 15, 2012, FPS decided to no longer use RAMP to help oversee its contract guard program.

Given FPS's challenges, you requested that we examine FPS's current efforts to conduct FSAs and oversee its contract guard workforce. This chapter examines the extent to which FPS is (1) completing risk assessments, (2) developing a tool to complete FSAs, and (3) managing its contract guard workforce. To examine the extent to which FPS is completing risk assessments and overseeing guards without RAMP, we reviewed, among other things, FPS's current FSA procedures and data on completed and planned FSAs for fiscal years 2010 to 2012. Specifically, we reviewed FPS's FSA data aggregated from its 11 regions to determine the extent of its FSA backlog. However, we could not determine the extent of the backlog because FPS's data contained a number of missing and incorrect values that made it unreliable. We also visited 3 of FPS's 11 regions and interviewed internal and external stakeholders including, among others, FPS, GSA, Department of

Veterans Affairs, the Federal Highway Administration, Immigration and Customs Enforcement, and guard companies. We selected these 3 regions based on the number of federal facilities in the region and their security levels, the number of contract guards in the region, and geographic dispersion. Our work is not generalizable to all FPS regions. To determine the status of FPS's efforts to develop an FSA tool, we reviewed, among other things, relevant project documents for MIST and federal physical security standards, such as DHS's *National Infrastructure Protection Plan's* (*NIPP*) risk management framework. We also interviewed FPS officials, representatives from Argonne National Laboratory who are responsible for developing MIST, and four risk management experts. We selected our four risk assessment experts from a list of individuals who participated in the Comptroller General's 2007 risk management forum.[3] We conducted this performance audit from July 2011 through August 2012 in accordance with generally accepted government auditing standards. Those standards require that we plan and perform the audit to obtain sufficient, appropriate evidence to provide a reasonable basis for our findings and conclusions based on our audit objectives. We believe that the evidence obtained provides a reasonable basis for our findings and conclusions based on our audit objectives. See appendix I for more detailed information on our scope and methodology.

BACKGROUND

To achieve its facility protection mission, in fiscal year 2012, FPS has a budget of $1.3 billion; over 1,200 full-time employees; and about 12,500 contract security guards. Contract guards are responsible for controlling access to federal facilities, screening access areas to prevent the introduction of weapons and explosives, enforcing property rules and regulations, detecting and reporting criminal acts, and responding to emergency situations involving facility safety and security. FPS relies on the fees it is authorized to charge federal tenant agencies in GSA-controlled facilities for its security services to fund its operations.[4] For example, FPS charges tenant agencies a basic security fee (currently $0.74 cents per square foot) to, among other things, conduct FSAs, monitor alarms and dispatch operations, and perform law enforcement activities. FPS's FSA process generally entails:

- gathering and reviewing facility information;
- conducting and recording interviews with tenant agencies;

- assessing threats, vulnerabilities, and consequences to facilities, employees, and the public; and
- recommending countermeasures to federal tenant agencies.

To carry out this process, FPS's long-term goal has been to develop a tool that aligns with DHS's *NIPP* risk-management framework and Interagency Security Committee (ISC) standards.[5] According to the *NIPP*, a risk assessment should assess threats, vulnerabilities, consequences, and recommend countermeasures, specifically:

- A *threat* assessment is the identification and evaluation of adverse events that can harm or damage an asset.
- A *vulnerability* assessment identifies weaknesses in physical structures, personal protection systems, processes, or other areas that may be exploited.
- A *consequence* assessment is the process of identifying or evaluating the potential or actual effects of an event, incident, or occurrence.

After these three assessments are completed, the information is used to determine whether a facility's risk is low, medium, or high. Additionally, the *NIPP* and ISC state that an agency's risk assessment methodology should be

- credible (or complete) as able to assess the threat, vulnerability, and consequences of specific acts;
- reproducible as able to produce similar or identical results when applied by various security professionals; and
- defensible as able to provide sufficient justification for deviations from the ISC defined security baseline.

In addition, as part of its FSA process, FPS also uses the ISC's *Facility Security Level Determination for Federal Facilities* to determine the facility security level (FSL). The ISC recommends that level I and II facilities be assessed every 5 years and level III and IV facilities every 3 years, and according to the ISC's criteria:

- A level I facility may be 10,000 or fewer square feet, have fewer than 100 employees, provide administrative or direct service activities, and have little to no public contact.

- A level II facility may be 100,000 or fewer square feet, have 250 or fewer employees, be readily identifiable as a federal facility, and provide district or statewide services.
- A level III facility may be 250,000 or fewer square feet, have 750 or fewer employees, be an agency's headquarters, and be located in an area of moderate crime.
- A level IV facility may exceed 250,000 square feet, have more than 750 employees, house national leadership, and be located in or near a popular tourist destination.

Since 2000, FPS has used three different tools to assess federal facilities and the assessment has varied, as shown in table 1.

Table 1. FPS's Past FSA Tools

Tools	Timeframe used	Description
Federal Security Risk Manager (FSRM)	2000 to November 2009	FSRM was a stand-alone computer assessment tool. With this tool, FPS's inspectors used a subjective approach to completing assessmentsand recommending countermeasures. However, FSRM did not assess risk according to the *NIPP's*risk management framework methodology orallow comprehensive analysis as the reports were not entered into a database.
Risk Assessment and Management Program (RAMP)	November 2009 to June 2010	RAMP was a Web-based riskassessment and guardmanagement tool. It was designed to calculate risks based on threat, vulnerability, and consequence using *NIPP's*risk management framework methodology. However, RAMP did not incorporate ISC's *Physical Security Criteria for Federal Facilities* because they were not finalized until after RAMP was developed.
FSA Calculator and Template	July 2010 to June 2011	The FSA calculator and template was an Excel spreadsheet and Word document that FPS's inspectors used to assess a facility's threat, vulnerability and consequence. After the assessments were completed, FPS planned to provide tenant agencies with a report with recommendedcountermeasures. The FSA calculator included RAMP's risk calculation methodology but also did not incorporate ISC's *Physical Security Criteria for Federal Facilities*or allow comprehensive analysis as the reports were not entered into a database.

Source: GAO analysis of FPS data.

FPS Does Not Currently Assess Risks at Federal Facilities, but Multiple Agencies Are Conducting Their Own Assessments

FPS Is Not Completing Risk Assessments

In the absence of RAMP, FPS currently is not assessing risk at the over 9,000 federal facilities under the custody and control of GSA in a manner consistent with federal standards such as NIPP's risk management framework, as FPS originally planned. As a result, FPS has accumulated a backlog of federal facilities that have not been assessed for several years. According to FPS data, more than 5,000 facilities were to be assessed in fiscal years 2010 through 2012. However, we were unable to determine the extent of the FSA backlog because we found FPS's FSA data to be unreliable. Specifically, our analysis of FPS's December 2011 assessment data showed that 9 percent—or nearly 800—of the approximately 9,000 facilities did not have a date for when the last FSA was completed. According to the NIPP, to be considered credible a risk assessment must specifically address the three components of risk: threat, vulnerability, and consequence. We have reported that timely and comprehensive risk assessments play a critical role in protecting federal facilities by helping decision makers identify and evaluate potential threats so that countermeasures can be implemented to help prevent or mitigate the facilities' vulnerabilities.[6]

Although FPS is not currently assessing risk at federal facilities, FPS officials stated that the agency is taking steps to ensure federal facilities are safe. According to FPS officials, its inspectors monitor the security posture of federal facilities by responding to incidents, testing countermeasures, and conducting guard post inspections. In addition, since September 2011, FPS's inspectors have been collecting information about federal facilities, such as location, purpose, agency contacts, and current countermeasures (e.g., perimeter security, access controls, and closed-circuit television systems). According to FPS officials, inspectors have collected information for more than 1,400 facilities that will be used as a starting point to complete FPS's fiscal year 2012 assessments. However, FPS officials acknowledged that this is not a credible risk assessment that addresses threat, vulnerability, and consequence consistent with NIPP's risk management framework. Moreover, several FPS inspectors told us that they received minimal training or guidance on how to collect this information and expressed concern that the facility information collected could become outdated by the time it is used to complete an FSA.

Multiple Federal Agencies Are Conducting Their Own Risk Assessments

We reported in February 2012 that multiple federal agencies have been expending additional resources to conduct their own risk assessments, in part because they have not been satisfied with FPS's past assessments.[7] These assessments are taking place even though according to FPS's Chief Financial Officer, FPS received $236 million in basic security fees from federal agencies to conduct FSAs and other security services in fiscal year 2011.[8] For example, an Internal Revenue Service (IRS) official said that IRS completed its own risk assessments based on concerns about risks unique to its mission for approximately 65 facilities that it also paid FPS to assess. A Federal Emergency Management Agency (FEMA) official stated that FEMA has assessed its own facilities for several years because of dissatisfaction with the facility security levels that FPS assigned to its facilities. Similarly, Environmental Protection Agency (EPA) officials said that EPA has conducted its own assessments based on concerns with the quality and thoroughness of FPS's assessments. EPA officials noted that the agency's assessments are conducted by teams of contractors and EPA employees, cost an estimated $6,000 each, and can take a few days to a week to complete. An official from the U.S. Army Corps of Engineers told us that it duplicates FPS's assessments at some of its regional facilities because the agency follows U.S. Army force protection regulations, rather than FPS's security requirements.

GSA is also expending additional resources to assess risk. We reported in October 2010 that GSA officials did not always receive timely FPS risk assessments for facilities GSA considered leasing.[9] GSA seeks to have these risk assessments completed before it takes possession of a property and leases it to tenant agencies. An inefficient risk assessment process for new lease projects can add costs for GSA and create problems for both GSA and tenant agencies. Therefore, GSA is updating a risk assessment tool that it began developing in 1998, but has not recently used, to better ensure that it has timely and comprehensive risk assessments. GSA officials told us that in the future they may use this tool for other physical security activities, such as conducting other types of risk assessments and determining security countermeasures for new facilities. However, as of June 2012, FPS has not coordinated with GSA and other federal agencies to reduce or prevent duplication of its assessments.

FPS Lacks Reliable FSA Data

In addition to not having a tool that allows it to conduct risk assessments, FPS does not have reliable FSA data, which has hampered the agency's ability to manage its FSA program. For example, as mentioned previously, we found that 9 percent—or nearly 800—of the approximately 9,000 facilities in FPS's dataset were missing a date for the completion of their last FSA, thus raising questions about whether facilities have been assessed as required.[10] Additionally, we found that FPS does not have reliable and timely information regarding when inspectors provided FSA reports to tenant agencies. This information is important because federal tenant agencies rely on these reports to allocate funding for new countermeasures.

We also found that FPS's reliance on its 11 regional offices to maintain FSA data has contributed to inconsistency among the regions. For example, each of the three regions we visited maintains FSA data in a different format. More specifically, each of the three regions collected similar information such as a facility's identifier and address, but they differed in how they tracked FSAs. For example, one region tracked the dates an FSA was submitted, reviewed, and completed. Another region tracked only the date the FSA was completed. Separately, another region used multiple spreadsheets to track FSAs. These inconsistencies among the regions make it difficult to understand whether FPS can manage its FSA program nationwide.

In March 2012, DHS's Inspector General (IG) also reported similar issues with FPS's data.[11] The IG found that FPS had not determined if any of the FSA data in RAMP were valid and thus needed to be preserved for future use. As a result, the IG stated that FPS risked incurring additional expenditures, including paying for the transfer of useless data or losing critical data, if it did not make a decision before June 2012, when its data maintenance contract expired. The IG recommended that FPS (1) identify the costs and benefits of two potential courses of action: maintaining the data in RAMP or transferring the data out of RAMP, and (2) review RAMP's data to determine what was critical and what should be saved. FPS concurred with this recommendation and plans to take action.

FPS Efforts to Develop a Risk Assessment Tool Are Evolving, but Challenges Remain

FPS Has Developed an Interim Vulnerability Assessment Tool

In September 2011, FPS signed an inter-agency agreement with Argonne National Laboratory for about $875,000 to develop MIST by June 30, 2012.[12] According to FPS's MIST documentation, MIST is an interim vulnerability assessment tool that FPS plans to use until it can develop a permanent solution to replace RAMP. According to FPS officials, among other things, MIST will enable the agency to begin aligning its FSA process with NIPP's risk management framework and ISC standards. In addition, according to FPS's MIST documentation, MIST will address key shortcomings identified with the RAMP development effort, including lack of inspector involvement, limited testing, and an inadequate training program.[13]

According to MIST project documents and FPS officials, among other things, MIST will also:

- allow FPS's inspectors to review and document a facility's security posture, current level of protection, and recommend countermeasures;
- provide FPS's inspectors with a standardized way for gathering and recording facility data; and
- allow FPS to compare a facility's existing countermeasures against the ISC countermeasure standards based on ISC's predefined threats to federal facilities (e.g., blast-resistant windows for a level IV facility) to create the facility's vulnerability report).[14]

In addition, according to FPS officials, after completing the MIST vulnerability assessment, inspectors will use additional threat information gathered outside of MIST by FPS's Threat Management Division and any local crime statistics to justify any deviation from the ISC-defined threat levels in generating a threat assessment report. FPS plans to issue the facility's threat and vulnerability reports along with any countermeasure recommendations to the federal tenant agencies.

FPS officials stated that MIST provides several potential improvements over its prior assessment tools: FSRM, RAMP, and the FSA calculator and template. For example, in contrast to FSRM, MIST will provide a more standardized and less subjective way of both collecting facility information and recommending countermeasures. Since MIST uses the ISC recommended countermeasures for defined threat scenarios for each facility security level, FPS officials believe that MIST will increase the likelihood that inspectors will produce credible FSAs. In contrast, the risk scores generated by RAMP and the FSA calculator and template were not linked to ISC standards. Unlike RAMP, MIST will use a limited amount of GSA facility data that can be edited by FPS inspectors where a correction is needed, according to FPS officials. The inability to edit data in RAMP was a contributing factor to its failure to produce credible FSAs.

According to FPS officials, on March 30, 2012, Argonne National Laboratory delivered MIST to FPS on time and within budget. FPS began training inspectors on MIST and about how to use the threat information obtained outside MIST in May 2012 and expects to complete the training by the end of September 2012. According to FPS officials, inspectors will be able to use MIST once they have completed training and a supervisor has determined, based on professional judgment, that the inspector is capable of using MIST. At that time, an inspector will be able to use MIST to assess level I or II facilities. According to FPS officials, once these assessments are approved, FPS will subsequently determine which level III and IV facilities the inspector may assess with MIST.

FPS Increased Its Use of Project Management Best Practices in Developing MIST

Considered Alternatives

FPS officials said the agency completed an alternatives analysis prior to selecting MIST. We were not able to confirm this because FPS did not document its analysis. According to industry standards, documenting an alternatives analysis is important because it allows agency officials to: revisit decision rationale when changes occur, reduce the subjectivity of the decision making process, and, provide a higher probability of selecting a solution that meets multiple stakeholders' demands.[15]

FPS officials mentioned two existing tools that were considered for an interim assessment tool: NPPD's Office of Infrastructure Protection's (IP) Infrastructure Survey Tool (IST) and DHS Science and Technology Directorate's (S&T) Integrated Rapid Visual Screening of Buildings (IRVS) tool. FPS officials said they became aware of a security survey conducted by IP for the February 2011 Super Bowl at Cowboys Stadium in Arlington,

Texas. Based on that survey, FPS reviewed the IST, which is used by IP to examine existing security countermeasures (which include physical and other protective measures) at critical infrastructure facilities, such as hydro-electric plants and commercial facilities, by comparing their existing countermeasures to those at similar facilities. According to IP officials, the IST does not calculate risk, estimate consequences, or recommend countermeasures. The IRVS is a risk assessment tool that assesses risk using threat, vulnerability, and consequence; that can be adapted to individual agency's needs; and that, according to an S&T official, was available to FPS at no cost. However, the Director of FPS decided that because of timeliness concerns and the opportunity to better share information within NPPD, FPS would develop a modified version of the IST to assess federal facilities until FPS could develop an FSA tool to replace RAMP.

Better Managed MIST's Requirements

In contrast to RAMP, FPS better managed MIST's requirements as we recommended in 2011.[16] Specifically, FPS's Director required that MIST be an FSA-exclusive tool and thus avoided changes in requirements that could have resulted in cost or schedule increases during development. Requirements serve as the basis for establishing agreement among users, developers, and customers and a shared understanding of the system being developed. Managing requirements entails managing the capabilities or conditions that a product is required to meet to satisfy an agreement or standard.

However, FPS did not obtain GSA or federal tenant agencies' input in developing MIST's requirements. We have reported that leading organizations generally include customer needs when developing programs. [17] Without this input, FPS's customers may not receive the information they need to make well-informed countermeasure decisions. FPS officials stated that they were considering getting feedback from GSA and federal tenant agencies.

Completed User Acceptance Testing

In March 2012, FPS completed user acceptance testing of MIST with some of its inspectors and supervisors, as we recommended in 2011.[18] User acceptance testing is conducted to ensure that a system meets contract requirements and performs satisfactorily for the user of the program—in this case, FPS's inspector workforce and their supervisors. The results of each test event need to be captured and used to ensure that any problems discovered are disclosed and corrected. We reported in 2009 that comprehensive testing that is effectively planned and scheduled can provide the basis for identifying key tasks and requirements. Testing can also better ensure that a system meets those specified requirements and functions as intended in an operational environment.[19]

According to FPS officials, user feedback on MIST was positive from the user acceptance test, and MIST produced the necessary output for FPS's FSA process. For example, the inspectors who were involved in the testing found the methodology understandable and credible and had no significant problems logging in and using MIST. FPS's testing identified the following problems: wireless connectivity issues at the testing location resulting in dropped connections and some users with older software encountering problems loading MIST onto their computers. FPS officials stated that they are taking steps to address these issues, such as updating older software.

MIST Has Limitations as an Assessment Tool

FPS has yet to decide what tool, if any, will replace MIST, which is an interim vulnerability assessment tool. According to FPS officials, the agency plans to use MIST for at least the next 18 months. Consequently, until FPS decides what tool, if any, will replace MIST or RAMP, it will continue to lack the ability to assess risk at federal facilities in a manner consistent with *NIPP*, as we previously mentioned. We also found the following limitations with MIST:

Assessing Consequence

FPS did not design MIST to estimate consequence, a critical component of a risk assessment. Assessing consequence is important because it combines vulnerability and threat information to evaluate the potential effects of an adverse event on a federal facility. For example, consequence information is used to determine whether a terrorist attack on a federal facility may result in the loss of human lives, incur economic costs beyond rebuilding the facility, or have an adverse impact on national security. Three of the four risk assessment experts we spoke with generally agreed that a tool that does not estimate consequences does not allow an agency to fully assess the risks to a federal facility. As a result, while FPS may be able to identify a facility's vulnerabilities to different threats using MIST, without consequence information, federal tenant agencies may not be able to make fully informed decisions on how to best allocate resources to protect facilities.

Both FPS and ISC officials stated that incorporating consequence information into an assessment tool is a complex task. FPS officials stated that they did not include consequence information in MIST's design as it would have introduced a new component that was not part of the IST and would have taken more time to develop, validate and test, and that any changes in threats would necessitate corresponding changes to the estimated consequences. For example, if new threats to federal facilities were identified, FPS would have to modify MIST's methodology to estimate the consequences and determine how those consequences could affect other previously identified threats. FPS officials do not know if this capability can be developed in the future, but they said that they are working with the ISC and S&T to explore the possibility. However, according to an S&T official, incorporating consequence is possible and S&T's current IRVS tool does estimate consequences.

Comparing Risk across Federal Facilities

FPS did not design MIST to compare risk or assessment results across federal facilities. Consequently, FPS does not have the ability to take a comprehensive approach to risk management across its portfolio of 9,000 facilities and recommending countermeasures to federal tenant agencies. Instead, FPS takes a facility-by-facility approach to risk management. Under this approach, FPS assumes that all facilities with the same security level have the same security risk, regardless of their location.[20] However, level I facilities typically face less risk because they are generally small store-front operations with a low volume of public contact, such as a small post office or Social Security Administration Office. In comparison, a level IV facility has a high volume of public contact and may contain high-risk law enforcement and intelligence agencies. We reported in 2010 that FPS's facility-by-facility approach to risk management provides limited assurance that the most critical risks at federal facilities across the country are being prioritized and mitigated.[21] FPS recognized the importance of having such a comprehensive approach to its FSA program when it developed RAMP and FPS officials stated that they may develop this capability for the next version of MIST.

Measuring Performance

FPS has not developed metrics to measure MIST's performance, such as feedback surveys from tenant agencies. Measuring performance allows organizations to track progress toward their goals and gives managers critical information on which to base decisions for improving their programs. We and other federal agencies have maintained that adequate and reliable performance measures are a necessary component of effective management.[22] We have also found that performance measures should provide agency managers with timely, action-oriented information in a format conducive to helping them make decisions that improve program performance, including decisions to adjust policies and priorities.[23] Without such metrics, FPS's ability to improve MIST will be hampered. FPS officials stated that they are planning to develop performance measures for MIST, but did not give a time frame for when they will do so.

FPS Faces Challenges in Overseeing Its Contract Guards

No Comprehensive System to Oversee Contract Guard Workforce

FPS does not have a comprehensive and reliable system to oversee its approximately 12,500 contract guards. In addition to conducting FSAs, FPS developed RAMP as a comprehensive system to help oversee two aspects of its contract guard program: (1) verifying that guards are trained and certified to be on post in federal facilities and (2) conducting guard post inspections.[24] However, FPS experienced difficulty with RAMP because the contract guard training and certification information in RAMP was not reliable.[25] Additionally, FPS faced challenges using RAMP to conduct post inspections. For example, FPS inspectors we interviewed stated they could not use RAMP to conduct post inspections because of difficulty connecting to RAMP's servers in remote areas and recorded post inspections disappearing from RAMP's record without explanation.

Although we reported some of these challenges in 2011, FPS did not stop using RAMP for guard oversight until June 2012. Consequently, it is now more difficult for FPS to verify that guards on post are trained and certified and that inspectors are conducting guard post inspections as required.

According to FPS officials, the agency decided to no longer use RAMP for these and other reasons, including the expiration of the RAMP operations and maintenance contract in June 2012 and FPS's decision to migrate data from RAMP. In the absence of RAMP, in June 2012, FPS decided to deploy an interim method to enable inspectors to record post inspections.

FPS officials said this capability is separate from MIST, does not include guard training and certification data, and will not have the ability to generate post inspection reports. In addition, FPS officials acknowledged that this method is not a comprehensive system for guard oversight.

No Independent Verification of Contract Guard Information

FPS does not independently verify the guard training and certification information provided by guard contractors. FPS currently requires its 33 guard contractors to maintain their own files containing guard training and certification information and began requiring them to submit a monthly report with this information to FPS's regions in July 2011.[26] To verify the guard companies' reports, FPS conducts monthly audits.

As part of its monthly audit process, FPS regional staff visits the contractor's office to select 10 percent of the contractor's guard files and check them against the reports guard companies send FPS each month.[27]

In addition, in October 2011, FPS undertook a month-long audit of every guard file for its contracts across its 11 regions. Similar to the monthly audits, regional officials explained that the "100 percent audit" included a review of the approximately 12,500 guard files for FPS's 110 contracts to verify that guards had up-to-date training and certification information.[28] According to an FPS official, the audit was FPS's first review of all of its contractors' guard files and provided a baseline for future nationwide audits.

FPS provided preliminary October 2011 data showing that 1,152 of the 12,274 guard files FPS reviewed at that time—9 percent—were deficient, meaning that they were missing one or more of the required certification document(s). However, FPS does not have a final report on the results of the nationwide audit that includes an explanation of why the files were deficient and whether deficiencies were resolved. FPS's monthly audits provide limited assurance that qualified guards are standing post, as FPS is verifying that the contractor-provided information matches the information in the contractor's files. We reported in 2010 that FPS's reliance on contractors to self-report guard training and certification information without a reliable tracking system of its own may have contributed to a situation in which a contractor allegedly falsified training information for its guards.[29]

In addition, officials at one FPS region told us they maintain a list of the files that have been audited previously to avoid reviewing the same files, but FPS has no way of ensuring that the same guard files are not repeatedly reviewed during the monthly audits, while others are never reviewed. In the place of RAMP, FPS plans to continue using its administrative audit process and the monthly contractor-provided information to verify that qualified contract guards are standing post in federal facilities.

CONCLUSION

FPS has taken some steps to improve its ability to assess risk at federal facilities but additional improvements are needed. Most notably, FPS has developed an interim vulnerability assessment tool that once implemented, may allow it to resume assessing federal facilities, which it has not done consistently for several years.

However, FPS's lack of progress in developing a risk assessment tool that meets federal physical security standards such as *NIPP's* risk management framework is problematic for several reasons.

First, FPS spent almost 4 years and $35 million dollars on RAMP and another $875,000 on MIST but still does not have a risk assessment tool that meets *NIPP's* risk management framework that can calculate risk using threat, vulnerability, and consequence information.

Second, without a risk assessment tool that can compare risks across its portfolio, FPS cannot provide assurance that the most critical risks at federal facilities are being prioritized and mitigated.

Third, some federal agencies are expending additional resources to conduct their own risk assessments in addition to paying FPS to complete them.

Fourth, federal tenant agencies do not have critical information needed to make risk-based decisions about how to upgrade the security of their facilities. Identifying ways to resolve these issues could greatly enhance FPS's efforts to assess risk at federal facilities and reduce duplication of effort, among other things.

We recognize that MIST is an interim tool and is not yet fully implemented; however, it has limitations that could affect FPS's ability to protect federal facilities and provide security services. FPS generally increased its use of our project management best practices, as we recommended, and we encourage it to continue to do so in any future development of a risk assessment tool.

However, FPS has not improved the accuracy and reliability of its FSA and contract guard data as it agreed to do in response to our previous recommendation. Given that FPS is still experiencing difficulties managing its FSA data, we reiterate the importance of this prior recommendation and encourage FPS to take action to address it.

Finally, FPS recently decided to not use RAMP to oversee its contract guards, but still does not have a comprehensive and reliable system to ensure that its approximately 12,500 contract guards have met training and certification requirements, and that FPS's guard post inspections are occurring in accordance with the agency's guidelines.

That FPS cannot ensure that its 33 contractors are providing accurate information on its guards is also problematic. Without a comprehensive and reliable system for contract guard oversight, FPS is relying primarily on information provided by guard companies. These issues raise important questions regarding the overall effectiveness of FPS's oversight of its contract guard workforce.

RECOMMENDATIONS FOR EXECUTIVE ACTION

Given the challenges that FPS faces in assessing risks to federal facilities and managing its contract guard workforce, we recommend that the Secretary of Homeland Security direct the Under Secretary of NPPD and the Director of FPS to take the following five actions:

- incorporate *NIPP's* risk management framework—specifically in calculating risk to include threat, vulnerability, and consequence information—in any permanent risk assessment tool;
- coordinate with GSA and other federal tenant agencies to reduce any unnecessary duplication in security assessments of facilities under the custody and control of GSA;
- address MIST's limitations (assessing consequence, comparing risk across federal facilities, and measuring performance) to better assess and mitigate risk at federal facilities until a permanent system is developed and implemented;
- develop and implement a new comprehensive and reliable system for contract guard oversight; and
- verify independently that FPS's contract guards are current on all training and certification requirements.

AGENCY COMMENTS

We provided a draft of this chapter to the Secretary of Homeland Security for review. DHS concurred with our recommendations. DHS also provided technical comments that we incorporated where appropriate.

Sincerely yours,
Mark L. Goldstein
Director, Physical Infrastructure Issues

APPENDIX I: SCOPE AND METHODOLOGY

To examine the extent to which FPS is completing risk assessments without RAMP, we reviewed FPS's current FSA procedures and data on completed and planned FSAs for fiscal years 2010 to 2012. Specifically, we reviewed FPS's FSA data aggregated from its 11 regions to determine the extent of FPS's FSA backlog. These data included the GSA facility identifier, address, city, state, zip code, FPS region, facility security level, date of the last FSA, and the date of the next scheduled FSA. However, we could not determine the extent of FPS's FSA backlog because FPS's data contained a number of missing and incorrect values that made it unreliable. We also visited 3 of FPS's 11 regions and interviewed regional managers, area commanders, and inspectors about how they are completing FSAs in the absence of RAMP. We selected these 3 regions based on the number of federal facilities in the region and their facility security levels, the number of contract guards in the region, and geographic dispersion. Our work is not generalizable to all FPS regions. We also interviewed FPS headquarters officials to understand how the agency is currently conducting FSAs. During our visits to the selected 3 FPS regions, we spoke with officials from the General Services Administration, Department of Veterans Affairs, the Federal Highway Administration, Immigration and Customs Enforcement, and United States Citizenship and Immigration Services to obtain their perspectives on FPS's assessment efforts. These agencies were selected because they are members of their facility security committees, which have responsibility for addressing security issues at their respective facilities and approving countermeasures recommended by FPS.

To determine the status of FPS's efforts to develop an FSA tool, we reviewed FPS's documents including: the interagency agreement, requirements plan, project plan, system test plan, and training plan for MIST. As applicable, we compared FPS's efforts to develop an FSA tool to DHS's *National Infrastructure Protection Plan's* (*NIPP*) risk management framework and the Interagency Security Committee's (ISC) standards, including the *Physical Security Criteria for Federal Facilities* and the *Facility Security Level Determination for Federal Facilities.*[1] We examined FPS's requirement and project documents to determine whether in developing MIST, FPS complied with selected GAO and industry project-management best practices, such as: conducting alternative analysis, managing requirements, and conducting user acceptance testing. These practices were selected because they are critical in developing information technology systems and we recommended in 2011 that FPS better manage its requirements and conduct user acceptance testing in developing future tools. We interviewed FPS headquarters and regional officials as well as inspectors, representatives from Argonne National Laboratory who are responsible for developing MIST, officials from NPPD's Office of Infrastructure Protection, and four risk management experts. We selected our four risk assessment experts from a list of individuals who participated in the Comptroller General's 2008 risk management forum.[2] We interviewed these experts to discuss FPS's efforts to assess risks to federal facilities and the benefits and challenges of a risk assessment.

To assess FPS's effort to manage its contract guard workforce, we reviewed FPS's guard oversight policies and procedures and RAMP's September 30, 2011, post inspection data.

During our visits to the selected three FPS regions, we interviewed FPS regional managers, area commanders, inspectors, three guard contractors, GSA, and other federal agencies about guard oversight. We also interviewed officials at FPS's headquarters.

We conducted this performance audit from July 2011 through August 2012 in accordance with generally accepted government auditing standards. Those standards require that we plan and perform the audit to obtain sufficient, appropriate evidence to provide a reasonable basis for our findings and conclusions based on our audit objectives. We believe that the evidence obtained provides a reasonable basis for our findings and conclusions based on our audit objectives.

End Notes

[1]GAO, Federal Protective Service: Actions Needed to Resolve Delays and Inadequate Oversight Issues with FPS's Risk Assessment and Management Program, GAO-11-705R (Washington, D.C.: July 15, 2011).

[2]GAO-11-705R.

[3]GAO, Highlights of a Forum: Strengthening the Use of Risk Management Principles in Homeland Security, GAO-08-627SP (Washington, D.C.: April 2008).

[4]40 U.S.C. 586; 41 C.F.R. § 102-85.35; Pub. L. No. 111-83, 123 Stat. 2142, 2156-57 (2009).

[5]The ISC is comprised of representatives from more than 50 federal agencies and departments, establishes standards and best practices for federal security professionals responsible for protecting non-military federal facilities in the U.S. FPS is a member agency of the Interagency Security Committee in the Department of Homeland Security, along with other federal agencies such as the General Services Administration, the Federal Aviation Administration, the Environmental Protection Agency, and other components within the Department of Homeland Security.

[6]GAO, Homeland Security, Greater Attention to Key Practices Would Improve the Federal Protective Service's Approach to Facility Protection, GAO-10-142 (Washington D.C.: Oct. 23, 2009).

[7]GAO, 2012 Annual Report: Opportunities to Reduce Duplication, Overlap and Fragmentation, Achieve Savings and Enhance Revenue, GAO-12-342SP (Washington, D.C.: February 2012).

[8]FPS currently charges tenant agencies in properties under GSA control a basic security fee of $0.74 per square foot per year for its security services including physical security and law enforcement activities as per 41 C.F.R. § 102-85.35.

[9]GAO-10-142.

[10]The ISC recommends that level I and II facilities be assessed every 5 years and level III and IV facilities be assessed every 3 years.

[11]Department of Homeland Security, Office of the Inspector General, FPS' Exercise of a Contract Option for the Risk Assessment and Management Program, OIG-12-67 (Washington, D.C.: March 2012).

[12]As of March 2012, FPS's total life cycle cost for MIST was estimated at $5 million.

[13]Federal Protective Service, MIST Integrated Systems Logistics Plan (Washington D.C., Mar. 27, 2012).

[14]The ISC has defined 31 different threats to federal facilities including vehicle-borne improvised explosive devices, workplace violence, and theft.

[15]Carnegie Mellon University Software Engineering Institute, Capability Maturity Model Integration for Acquisition, Version 1.2 (November 2007).

[16] GAO-11-705R.

[17]GAO, Geostationary Operational Environmental Satellites: Improvements Needed in Continuity Planning and Involvement of Key Users, GAO-10-799 (Washington, D.C.: Sept.1, 2010).

[18]GAO-11-705R.

[19]GAO, Information Technology: Census Bureau Testing of 2010 Decennial Systems Can Be Strengthened, GAO-09-414T (Washington, D.C.: Mar. 5, 2009).

[20] GAO-10-142.

[21]GAO, Homeland Security: Addressing Weaknesses with Facility Security Committees Would Enhance Protection of Federal Facilities, GAO-10-901 (Washington, D.C.: Aug. 5, 2010).

[22]GAO, Homeland Security: The Federal Protective Service Faces Several Challenges That Hamper its Ability to Protect Federal Facilities, GAO-08-683 (Washington, D.C.: June 11, 2008).

[23]GAO-08-683.

[24]FPS's inspection requirement for FSL I and II facilities is two annual inspections of all posts, all shifts. The inspection requirement for FSL III facilities is biweekly inspections of two posts, any shift, and for FSL IV, weekly inspections of two posts, any shift.

[25]A post is a guard's area of responsibility in a federal facility.

[26]For example, guard training and certifications include firearms qualification, cardiopulmonary resuscitation, First Aid, baton certification, and x-ray and magnetometer training.

[27]FPS now relies on guard contractors to keep accurate guard certification records. Each month, regional personnel are required to review 10 percent of the contractors' guard certification files to verify that the information is current and matches the monthly guard certification spreadsheet FPS receives from the contractors. According to FPS policy, if regional personnel identify deficiencies, such as expired certification documentation, in 40 percent of the files reviewed, they are to initiate an audit of 100 percent of the company's files.

[28] A guard company may have more than one contract with FPS.

[29]GAO, Homeland Security: Federal Protective Service's Contract Guard Program Requires More Oversight and Reassessment of Use of Contract Guards, GAO-10-341 (Washington, D.C.: Apr. 13, 2010).

End Notes for Appendix I

[1]The ISC is comprised of representatives from more than 50 federal agencies and departments, establishes standards and best practices for federal security professionals responsible for protecting non-military federal facilities in the U.S. FPS is a member agency of the Interagency Security Committee in the Department of Homeland Security, along with other federal agencies such as the General Services Administration, the Federal Aviation Administration, the Environmental Protection Agency, and other departments within the Department of Homeland Security.

[2]GAO, Highlights of a Forum: Strengthening the Use of Risk Management Principles in Homeland Security, GAO-08-627SP (Washington, D.C.: April 2008).

In: Federal Protective Service
Editor: Emilius M. Zamaco

ISBN: 978-1-62417-205-2

Chapter 4

TESTIMONY OF LEONARD E. PATTERSON, DIRECTOR, FEDERAL PROTECTIVE SERVICE, U.S. DEPARTMENT OF HOMELAND SECURITY. HEARING ON "SECURING FEDERAL FACILITIES: AN EXAMINATION OF FPS PROGRESS IN IMPROVING OVERSIGHT AND ASSESSING RISK"*

Thank you Chairman Lungren, Ranking Member Clarke, and the distinguished members of the Subcommittee. My name is Eric Patterson, and I am the Director of the Federal Protective Service (FPS) within the Department of Homeland Security's (DHS) National Protection and Programs Directorate (NPPD).

I am honored to appear before you today to discuss NPPD/FPS's progress in utilizing key protection and risk management practices such as allocation of resources, leveraging technology, and enhancing information sharing and coordination.

The GAO has raised several areas that have historically represented challenges for FPS including:

---

* This is an edited, reformatted and augmented version of testimony given on July 24, 2012 before the House Committee on Homeland Security, Subcommittee on Cybersecurity, Infrastructure Protection, and Security Technologies.

1. Absence of a risk management program;
2. Addressing key human capital issues through a strategic human capital plan;
3. Contract Guard workforce management and oversight; and
4. Need for a review of FPS's fee design

Today's hearing is an opportunity to address the progress FPS has made during the past year in working to address these challenges, and to also provide information on the topics addressed in GAO's new chapter related to risk assessment and Protective Security Officer (PSO) program management and oversight.

FPS Background

FPS's mission is to protect more than 9,000 Federal buildings and the 1.4 million Federal employees and visitors who occupy them throughout the country every day by leveraging the intelligence and information resources of its network of public and private sector partners. Specifically, FPS executes its mission by providing proactive law enforcement, investigation and protective intelligence and information sharing services, incident response, security planning, and stakeholder engagement. Prior to its transfer to NPPD in 2009, FPS was organized under Immigration and Customs Enforcement and prior to that, under the General Services Administration (GSA).

Part of our core mission is to assess the threat picture for the Government Facilities Sector (GFS) and share that information with stakeholders as appropriate. For example, FPS leverages the Homeland Security Information Network (HSIN), a secure, trusted web-based portal to share information with our more than 900 government and industry partners. One of the recent information-sharing initiatives FPS has implemented to assist in the protection of facilities and their occupants is the Federal Facility Threat Picture (FFTP), which is an unclassified assessment of the current known threats to the facilities FPS protects. Produced quarterly, the FFTP supports the threat component of a Federal Security Assessments (FSA) and informs our stakeholders of potential threats to government facilities. The FFTP focuses on the threats posed by a variety of actors that may seek to attack or exploit elements of the GFS. The information used in the FFTP comes from intelligence and law enforcement community reporting.

During Fiscal Year (FY) 2011, FPS:

- Investigated and mitigated more than 1,300 threats and assaults directed towards Federal facilities and their occupants;
- Disseminated 331 threat and intelligence-based products to our stakeholders, 142 of which were FPS-produced;
- Conducted 81,125 post inspections;
- Interdicted more than 680,000 weapons/prohibited items including knives, brass knuckles, pepper spray, and other items that could be used as weapons or are contraband such as illegal drugs, at Federal facility entrances during routine checks;
- Made 1,975 arrests;
- Responded to 53,000 incidents involving people or property; and
- Conducted more than 1,800 high-visibility operations under Operation Shield and 150 risk-based Covert Test operations, ensuring the protection of Federal buildings and infrastructure.

FPS Is Developing a Risk Management Program

In terms of a risk management program, FPS's operational activities are organized by the National Infrastructure Protection Plan's (NIPP) Risk Management Framework, which calls for the following steps: Set Security Goals, Identify Assets and Functions, Assess Risks, Prioritize, Implement Protective Programs, and Measure Effectiveness. One area of recent significant progress related to risk assessment and the implementation of a risk management program is the ongoing implementation of FPS's solution for conducting FSAs using an automated assessment tool. In May 2011, the decision was made to cease development of the legacy application known as the Risk Assessment and Management Program (RAMP) and to pursue a standalone assessment tool, in order to provide completed FSAs to customers. That decision has since been affirmed by the Department's Office of Inspector General (OIG).

In the interim period, our employees have continued their daily interactions with tenant agencies and oversight of facility security. Our personnel have been completing Pre-Modified Infrastructure Survey Tool (MIST) worksheets to enable complete FSA reports, and are constantly assessing risks to Federal facilities. Specifically, the pre-MIST worksheet allows the inspector to collect key information that will be populated into

MIST and used in generating a final FSA report. Such data includes facility information, vulnerability assessments, and existing protective measures.

After consideration of several alternatives, FPS partnered with NPPD's Office of Infrastructure Protection (IP) to leverage a proven assessment methodology called the Infrastructure Survey Tool (IST). In October 2011, NPPD issued a task order to Argonne National Laboratory (ANL) through the Department of Energy to modify the existing Link Encrypted Network System (LENS) and IST for FPS use to conduct FSAs. Because this project leveraged existing tools and had limited resources and time constraints, the acquisition lifecycle was tailored to meet delivery deadlines.

I am pleased to note that in its draft chapter, GAO noted FPS's use of project management principles in the development of MIST. Throughout the project, the MIST Users Working Group has remained engaged to ensure user involvement in the process. User feedback from field testing was uniformly positive about MIST and the FPS Gateway, confirming suitability to support the FPS mission. The MIST and FPS Gateway development efforts were completed on schedule, with ANL delivering the system to the Government on March 30, 2012. In April 2012, and the decision was made to proceed and deploy MIST. It is important to note that throughout the development and testing of MIST, field employees and our union were involved and actively participated as subject matter experts in the process.

FPS developed and is currently implementing a distance learning-based training program for each MIST user, as GAO commended in its draft chapter. Supervisors completed this training in April 2012 and Inspectors began their virtual training in May 2012, with completion of all training anticipated for late September 2012. This provides a hands-on learning environment for our inspectors; they will receive virtual instruction as they use the tool in the learning environment. Once an Inspector completes the training and successfully briefs his or her supervisor on a completed FSA, that Inspector will be able to proceed with conducting FSAs and reporting the results to a Facility Security Committee.

In leveraging existing technology in developing MIST, FPS was able to incorporate the ability to illustrate the impact of alternative countermeasures on a particular vulnerability. MIST will also show how a facility is or is not meeting the baseline level of protection for its Facility Security Level as set forth in the ISC's Physical Security Criteria for Federal Facilities standard and the ISC's Design Basis Threat report. This will lead to a more informed and better dialogue with tenants and Facility Security committees as FSA results are discussed and alternatives are explored. Additionally, FPS recently disseminated guidance nationwide on the commencement of the use of MIST to generate FSAs upon completion of inspector training. The anticipated results of the use of MIST are consistent assessment results nationwide and informed decision-making regarding security investments on the part of tenant agencies.

FPS IS ADDRESSING KEY HUMAN CAPITAL ISSUES THROUGH DEVELOPMENT OF A STRATEGIC HUMAN CAPITAL PLAN

In order to ensure that human resource requirements are aligned appropriately with FPS's overall mission, a Strategic Human Capital Plan is being developed in conjunction with NPPD's Human Capital Office. We are working to finalize the document; we intend to provide the plan and brief the Committee when it is finalized.

FPS IS WORKING TO IMPROVE ITS PROTECTIVE SECURITY OFFICER MANAGEMENT AND OVERSIGHT

FPS is working to improve management and oversight of our over 13,000 Protective Security Officer (PSO) force. We have reviewed our operations nationwide and have taken steps at the national program level to ensure that performances under contracts are advantageous to the Government. We are actively working to implement the recommendations resulting from GAO and OIG reviews across the organization. Additionally, an Integrated Project Team (IPT) conducted a comprehensive review of how FPS resources the PSO oversight function and our current oversight policy.

FPS is also working with DHS's Science and Technology Directorate to develop a system for contract guard oversight and explore means of leveraging technology to ensure effective oversight of PSOs, such as automated tracking of guard post staff levels and PSO possession of the necessary credentials to stand post. Additionally, our training team is working closely with industry and Federal partners in developing a more effective training strategy for our PSOs.

FPS IS EXAMINING ITS FEE STRUCTURE IN ORDER TO REVIEW CURRENT FEE DESIGN

FPS operates through fee-based funding revenue, which is calculated based on the Federal-facility tenant's square footage of occupancy and on the collection of services associated with the provisioning of reimbursable protective countermeasures. This fee-based financial structure is unique among Federal law-enforcement agencies and requires a greater degree of understanding internal operations to ensure it is properly aligned with FPS's costs.

To address this challenge, FPS is implementing a two-pronged strategy to better understand its activities and costs and recommend options for a new revenue structure. In January 2012, FPS collaborated with the Department's Systems Engineering and Design Institute (SEDI), a Federally Funded Research and Development Center managed by the DHS Science and Technology Directorate, to produce a full mapping of FPS activities and then align them with costs. That work will be used to produce Activity-Based Cost (ABC) models for FPS. Both of these efforts are designed to result in a more efficient revenue structure for FPS and greater transparency in security costs for FPS stakeholders.

CONCLUSION

Thank you again for the opportunity to provide you with an update on the progress FPS is making on a number of fronts. FPS aspires to be an exemplary law enforcement and strategic critical-infrastructure protection organization. This is a vision uniformly shared by FPS leadership and operational staff, both at headquarters and in the field. I would be happy to answer any questions you might have.

INDEX

D

E

F

G

H

I

J

L

M

N

O

T

U

V

W